Why the Social Sciences Matter

Why the Social Sciences Matter

Edited by

Jonathan Michie
President, Kellogg College, University of Oxford, UK

and

Cary L. Cooper
Distinguished Professor of Organizational Psychology and Health,
Lancaster University, UK

palgrave
macmillan

First published 2015 by
PALGRAVE MACMILLAN

Palgrave Macmillan in the UK is an imprint of Macmillan Publishers Limited,
registered in England, company number 785998, of Houndmills, Basingstoke,
Hampshire RG21 6XS.

Palgrave Macmillan in the US is a division of St Martin's Press LLC,
175 Fifth Avenue, New York, NY 10010.

Palgrave Macmillan is the global academic imprint of the above companies
and has companies and representatives throughout the world.

Palgrave® and Macmillan® are registered trademarks in the United States,
the United Kingdom, Europe and other countries.

ISBN: 978–1–137–26990–4 hardback
ISBN: 978–1–137–26991–1 paperback

This book is printed on paper suitable for recycling and made from fully
managed and sustained forest sources. Logging, pulping and manufacturing
processes are expected to conform to the environmental regulations of the
country of origin.

A catalogue record for this book is available from the British Library.

A catalog record for this book is available from the Library of Congress.

Contents

Foreword

The post-war American social scientist, C. Wright Mills, once wrote of the obligation to turn 'private troubles' into 'public issues'. He was referring specifically to the discipline of sociology, but this aphorism applies to the social sciences more generally.

It seems strange, therefore, that at the present time social science is under such virulent political attack in the United States, where federal research funding has to meet extraordinarily restrictive – and politically motivated – criteria relating to short-term utility and national security. And this is aimed at the most prestigious and influential social science research community in the world.

These things go in cycles (Wright Mills himself suffered from the McCarthyite witch hunts of the 1950s) and the counter-veiling civilising impulses of American society will doubtless see the cycle turn once more. But in the aftermath of the post-2008 global financial crisis, it is scarcely surprising that segments of American public and political opinion have fastened upon the economics discipline in particular, and the social sciences more generally, as a convenient scapegoat.

One can observe distant, and muted, echoes of this in comments in the press and media in this country – and of course we witnessed the attempt to abolish the (then) Social Science Research Council in the 1980s. But no one seriously proposes the abolition of the Economic and Social Research Council today – and if that signifies a kind of progress, it is equally the case that the public acceptance of the importance of the social sciences cannot be taken for granted.

In the UK, the history of many social science disciplines emerges out of Wright Mills' distinction. Private troubles became public issues by virtue of detailed empirical enquiry providing both the evidence for public reform and a realisation that there were causes of private troubles which lay beyond the purview of the individual either to understand or remedy them. This was the classic Fabian agenda. Although it was never quite as simple as this, it at least provided a role for social science which remains a component of public discourse up until the present.

So it does no harm, from time to time, to reassert the importance of social science in building a civilised and civilising society. If we look at the so-called research 'grand challenges' of RCUK or the European

Commission – climate change, sustainable resource utilisation, food security, public health, and so on – can anyone seriously claim that they do not have a social science dimension? And neither are we referring to consequences just of scientific and technological innovation, but also of the socio-economic conditions which foster, or inhibit, such change.

Social science is about evidence, but not only this. It is also about ideas, insight, understanding and, crucially, debate when there are no simple solutions to our collective private troubles. As a community we should not be over-defensive but it behoves us to demonstrate our value. The following chapters do just that.

Howard Newby
President of Academy of Social Sciences

List of Contributors

Chloe Campbell is a Research Fellow at the Psychoanalysis Unit, University College London, UK.

Cary L. Cooper, CBE, is the Distinguished Professor of Organizational Psychology and Health at the Management School, Lancaster University, UK.

Michael A. Crang is Professor in the Department of Geography, Durham University, UK.

Stuart Croft is Professor and Pro Vice Chancellor for Research (Arts and Social Sciences) at the University of Warwick, UK.

Pasco Fearon is Professor of Developmental Psychopathology in the Research Department of Clinical, Educational and Health Psychology, University College London, UK.

Robert J. Gatchel, PhD, ABPP is Nancy P. & John G. Penson Endowed Professor of Clinical Health Psychology and Distinguished Professor of Psychology at the University of Texas at Arlington, USA.

Nicky Gregson is Professor of Human Geography in the Department of Geography, Durham University, UK.

Oz Hassan is Assistant Professor in US National Security in the Department of Politics and International Studies, University of Warwick, UK.

Rod Hick is a Lecturer in Social Policy at Cardiff University, UK.

Mike Hough is Professor of Criminal Policy at the School of Law, Birkbeck, University of London, UK and Associate Director of the Institute for Criminal Policy Research, UK.

Mavis Maclean, CBE, is Senior Research Associate in the Department of Social Policy and Intervention, University of Oxford, UK.

Cathy McIlwaine is Professor of Geography at the School of Geography, Queen Mary, University of London, UK.

Jonathan Michie is Professor of Innovation and Knowledge Exchange and President of Kellogg College, University of Oxford, UK.

Lynne Murray is Professor of Developmental Psychopathology at the School of Psychology and Clinical Language Sciences, University of Reading, UK.

James Campbell Quick holds the John and Judy Goolsby – Jacqualyn A. Fouse Endowed Chair at the University of Texas at Arlington, USA and is Distinguished Visiting Scholar at the Management School, Lancaster University, UK.

Ceridwen Roberts, OBE, is Senior Research Fellow in the Department of Social Policy and Intervention, University of Oxford, UK.

Camilla Toulmin is Director of the International Institute for Environment and Development, UK.

John Urry is Distinguished Professor in the Department of Sociology, Lancaster University, UK.

Introduction

Jonathan Michie and Cary L. Cooper

Just opening a newspaper serves as a reminder of the problems with which society is beset – problems big and small. Banks are in trouble; climate change threatens; the worsening headache of caring for increasingly frail elderly parents; householders being charged for rubbish disposal; youngsters on the rampage; phone hacking; furious rows about 'genetic engineering'; the unsayable about immigration haunting policies on unemployment; another family grieving for their dead soldier child who did not come home from an expensive war. Public issues and private troubles are as interlaced as ever. And, big or small, problems need solutions and solutions need to be based on accurate and suitable information, and on a proper understanding of the issues involved.

This volume takes a considered look at a selection of problems such as these. In the process it showcases contemporary work in the social sciences. It consists of eleven specially commissioned essays on topics of prime concern at the beginning of the twentieth century to all in the United Kingdom (UK) – and globally. All tackle difficult questions and involve problems that are complex and hard to solve. Above all, each essay in its own way illuminates why having an understanding of a social scientific 'take' on the topic in question provides a grasp that would not be provided by any other 'knowledge producer'; in other words, the chapters make plain what is distinctive and thus invaluable about whichever social science is being presented, and why social science needs to be included if these issues are to be properly understood, and appropriate policies developed.

What sets this book apart from others is that all its chapters are written by experienced social scientists who work at the cutting edge of their respective research fields. Of course there is a wide array of writing on each topic by journalists or politicians, activists and self-appointed

1

commentators, from think-tanks and non-governmental organization's (NGOs), along with a steady stream of official documents, statistics and reports from governments and others in the UK, the EU and further afield. What sets this collection apart from those other contributions is that it presents the upshot of systematic scholarly investigation. Certainly the best of those others draw, from time to time, on this scholarship, as well as adding to it: the boundaries are permeable. Here, though, is the chance to read that scholarship first hand. Thus scholarly virtues are made apparent. So the book displays how the considerable complexity of each topic is grasped intellectually, and sets out the way research is methodically undertaken, with the results analysed and presented even-handedly. It confirms the good reasons for pursuing evidence-based policy.

The book originates in the UK – enjoying the active support of the Academy of Social Sciences. But it is not exclusively UK focused. This is partly because it recognises that problems of the sort referred to above are not exclusive to the UK. Partly too because the contemporary world and its concerns are heavily interconnected. Also, most importantly, scholarship transcends national borders. The various contributors have looked beyond the shores of the British Isles, not just to compare and contrast but also to analyse transnational consequences, and cross border effects and implications.

This volume is timely in analysing pressing problems, and is also timely in helping make the case regarding the indispensability of the social sciences. Social scientists may be forgiven for being troubled by UK trends in funding undergraduate training and the 'impact' agenda in research. It would be curious for a group not to react with alarm when a government introduces policies affecting directly – and seemingly adversely – their interests. So as a side effect, albeit a most important one, this volume helps demonstrate the value of attempting to understand social life now and into the future. It builds on and complements the series of lectures and publications launched by the Academy of Social Sciences in 2010 under the heading of 'Making the Case for the Social Sciences'.

Overview

In Chapter 1, on Social Science, Parenting and Child Development, Pasco Fearon, Chloe Campbell and Lynne Murray analyse the study of children's development, arguing that it represents a key area in which social science can make a vital contribution to scientific knowledge, clinical practice, social policy and wider society. They give examples that serve to illustrate this kind of approach, and the ways in which child development research is enriching our understanding of the importance of parenting for children's healthy social, cognitive and emotional development. Specifically, they provide an overview of research on attachment and postnatal depression, and examine what is currently known about the effects of these on child development, the social interaction mechanisms that convey risk, and the ways in which these two areas of research have been used to develop intervention programmes and reshape social and health policy.

In Chapter 2, on Health and Wellbeing, James Campbell Quick, Robert Gatchel and Cary L. Cooper draw on the practice of preventive health management, on established learning principles, on behavioural and social sciences, and on emerging positive practices. The Chapter opens with a discussion of the major preventable health risks that can undermine health and wellbeing. By building on strengths, guarding against risks and compensating for vulnerabilities, health and wellbeing can be enhanced. The chapter next shows how established learning principles offer powerful and positive ways to advance health and wellbeing through the behavioural and social sciences. Three specific learning pathways are through: classical conditioning; operant conditioning; and observational learning, or modelling. Attention is given to the environmental context within which learning occurs, especially the work environment. The chapter concludes with a section on positive psychological and organisational wellbeing as enhanced through the behavioural and social sciences.

In Chapter 3, on Climate Change and Society, John Urry demonstrates just why society is so important in analysing the nature of climate change. And because society is important, so the social sciences need to be brought directly into examining the causes of change – and the likely ways in which climate change might be mitigated. So far the social sciences have played a minor role by comparison with the physical sciences and economics. The chapter explores the many ways in which society is central to changing climates – and the mitigation of such changes. This is not just a question of changing what individuals do, but of changing whole *systems* of economic, technological and social practice, which presuppose patterns of social life which become embedded and relatively unchanging for long periods. These high carbon systems have locked into social life – and breaking these lock-ins is particularly challenging. It is the need to understand those systems that make the social sciences key to future analysis and policy development.

In Chapter 4, on Waste, Resource Recovery and Labour: Recycling Economies in the EU, Nicky Gregson and Mike Crang show that the social sciences are critical to the challenge of turning wastes into resources via materials recovery and recycling. For wastes to become resources they have to become products, bought and sold in markets. Economics, and the economic geographies of manufacturing, matter when it comes to deciding what can be done, where, with particular wastes. The chapter draws on original research in three sectors – dry recyclables, ship recycling and textiles recovery – to show the difficulties that confront the policy goal of turning the European Union (EU) into a recycling economy. It shows that recycling in the EU is positioned in the secondary labour market; that these jobs are characterised by itinerant migrant labour, often from East and Central Europe (ECE); and that low-grade products require global recycling networks to realise value.

In Chapter 5, on Poverty and Inequality, Rod Hick argues that the problems of poverty and inequality remain important public – and political – concerns. Hick examines the literature on poverty and inequality, drawing on two cases. The first involves some of the early analysis of poverty in the UK in the nineteenth century, and Hick shows how the discussion of contemporary poverty analysis has built on this earlier tradition; the second is the recent publication, and subsequent debate surrounding *The Spirit Level*, by Richard Wilkinson and Kate Pickett. The chapter argues that, as partially observable phenomena, there is much about poverty and inequality which is readily accessible to the 'non-expert'. The contribution which the social sciences can make is, Hick argues, in terms of the rigour to which the social sciences can aspire in understanding the social world.

In Chapter 6, on The Economy, Financial Stability and Sustainable Growth, Jonathan Michie argues that social science has a major and important role to play in analysing the nature and functioning of the economy. Many of the major economic issues are linked inseparably to other areas of social science research and interest, such as inequality of income and wealth, and the effects of this on society. Indeed, many of the 'classic' texts analysing the economy – from Adam Smith's *Wealth of Nations*, to Marx's *Capital* and Keynes's *General Theory* – touched on a range of issues beyond the narrowly economic. One such discussion today relates to whether measures of economic growth or progress need to take account of broader aspects than previously thought necessary, whether in welfare or sustainability. The chapter discusses these issues, including with reference to the 2007–2008 global financial crisis, globalisation, and the need to foster a more corporately diverse economy and financial services sector. In terms of environmental sustainability, the importance of complexity is stressed; it is argued that the economy cannot be understood adequately through 'marginal' analysis, and that instead systems theory and interdisciplinary approaches are called for.

In Chapter 7, on What Can the Social Sciences Bring to an Understanding of Food Security?, Camilla Toulmin argues that food security and politics have been closely linked since ancient times. Today, questions of food security focus on how to increase total supply for a possible 9–10 billion people by 2050. It is the job of social scientists to ask difficult questions, place food and agricultural systems in the bigger picture, and ask why, in a world of plenty, a billion people across the world still go hungry. In the last 20 years, social science has helped turn received wisdom on its head, by putting local people, their knowledge, insights and priorities to the fore. This has shown the importance of 'how' food is produced as well as 'how much'. Social science must also offer answers to the big public policy issues of the day, alongside the natural sciences, and address issues of politics, power and interests. Many of the most interesting food security questions cross the biophysical and socio-economic disciplines, and demand a joint approach if a just and sustainable solution is to be found.

In Chapter 8, on Family, Marriage and Divorce, Mavis Maclean and Ceridwen Roberts applaud the contribution of accessible and high quality demographic data on the family, but argue for care in interpretation. The chapter debates the value of the contribution of expert social scientists in not only using the data to respond to the questions currently facing policy makers, but also drawing on these data to develop an understanding of emerging issues, and of the questions which policy makers will need to face in the future.

In Chapter 9, on Crime, Policing and Compliance with the Law, Mike Hough argues that social scientific research has made a very substantial contribution to specialist academic understanding of crime and its control. The chapter sketches out the contribution that has been made in three areas: our understanding of crime trends; our knowledge of policing and its effects of crime; and the factors that encourage people to comply with the law. The ways in which practitioners and academics think about these issues has been transformed over the last half-century, and social scientific research is a significant factor in achieving this transformation. However, the same research has achieved a much more tenuous hold on political and public discourse about crime, and the chapter concludes with a discussion of the reasons for this, and offers some thoughts on how social science should aim to extend its reach into highly politicised issues such as 'law and order'.

In Chapter 10, on Understanding the Arab Spring, Stuart Croft and Oz Hassan demonstrate how the social sciences have contributed to our understanding of why the Arab Spring took place. Moving beyond predominant and immediate narratives, it argues that what the social sciences teach us is a way of analysing unfolding security situations at multiple levels. The chapter demonstrates the roles that demographics, technology, pluralism, political economy, military decisions, historical contexts and global effects played in contributing to the revolutions taking place across the Middle East and North Africa.

Finally, in Chapter 11, on International Migration, Cathy McIlwaine argues that as one of the most important yet most contentious phenomena of our times, international migration has not only grown, but has become increasingly complex. With important temporal and spatial variations in the dynamics and delineations of international migration, it is appropriate to approach understandings of it from a wide range of disciplinary perspectives in order to capture the different scales of analysis, methodologies and theoretical standpoints. The multi-disciplinary nature of the social sciences makes them ideally placed to understand the complexities of international migration movements in ways that other sets of disciplines are unable. Crucially, the social sciences are also central in providing robust and independent research that can challenge the often negative public perceptions of international migration in both destination and source countries.

Conclusion

These contributions make powerful cases across an important array of pressing issues. None of the authors seek to promote the importance of social sciences as against other areas of research or enquiry – on the contrary, the clear message is that to understand the major issues of the day requires genuinely interdisciplinary approaches, and this has to apply across the whole range of academic disciplines as appropriate – certainly not just restricted to the social sciences alone. Thus, to understand climate change of course requires the natural sciences; but equally, it requires an understanding of the worlds of management and business, of consumer behaviour, and of public policy behaviour – all of which require the social sciences. Neither are the authors complacent about the state of social science itself. The need for improved interdisciplinary working applies within the social sciences as well as between social science and other disciplines. There are thus lessons for us all to learn, in order to understand better the processes at work behind the various policy challenges that society faces, and to then build on this understanding to develop appropriate policy responses. The clear message is that deepening our understanding of many of the major issues facing us and future generations, improving the quality of public discussion, and assisting in the development of effective policy, depends crucially on the social sciences. The social sciences themselves need to rise to this challenge. The authors in their contributions to this book have certainly heeded that call.

1

Social Science, Parenting and Child Development

Pasco Fearon, Chloe Campbell and Lynne Murray

Introduction

Securing and promoting the welfare and healthy development of children should be one of the fundamental priorities, and challenges, for all societies. Despite notable progressive national initiatives, global policy statements, aid programmes and grassroots campaigns that are focused on children's health and wellbeing, many children continue to be exposed to major impediments to their optimal development, including poor parental care, outright abuse and neglect, domestic violence, poorly managed parental mental illness, displacement, poverty and lack of access to high quality educational, intellectual and creative opportunities (Walker et al., 2011). Very cogent arguments have been made that intervention in early child development can reap disproportionately higher returns in social and economic benefits than interventions focused on later periods of the lifespan (Heckman, 2008). Few would deny that the prevention of mental health problems, psychological distress, educational dropout and underachievement, unemployment and social maladjustment is better than cure. However, effective prevention requires a systematic understanding of the developmental mechanisms of maladjustment and a rigorous analysis of what interventions work, and for whom. These, in turn, hinge on critical analysis, rigorous measurement, good theory and carefully executed research.

Social science has a vital role to play in this arena: in systematically documenting the experiences and outcomes of children; in understanding the proximal (those impinging directly on the child) and distal (those contextual factors supporting or maintaining proximal effects on the child) mechanisms that affect children's development; and in developing and evaluating interventions and policies for changing children's lives for the better.

The ways in which social science has already contributed to children's health and development are too numerous to cover in a short chapter. Instead, we focus on just two examples by way of illustration to show the rigorous, conceptually coherent and principled approaches that social scientists offer for advancing our understanding of child development and delivering new and better ways of promoting children's outcomes across the globe. Inevitably, we are not able to cover the many crucial areas of social science that have made just as much of a contribution as the examples we have chosen. Social science, for example, has produced vital research on typically developing children's language acquisition, peer relations, intellectual development, educational attainment and learning, citizenship and moral development, as well as the development and needs of children with disabilities. All of these domains of research are rich in scientific data and theory, and have been translated into effective social and educational interventions. Social scientists have also initiated some of the most important large-scale surveys that have given the public and policy-makers vital insights into the health, experiences, needs and opinions of children in our societies, such as the UK Household Longitudinal Study (Bradshaw, Keung, Rees and Goswami, 2011), the Avon Longitudinal Study of Parents and Children (Golding, 1990) and the Millennium Cohort Study (e.g., Sabates and Dex, 2012) in the UK or the NICHD Study of Early Childcare and Youth Development (NICHD, 2005) and the Early Childhood Longitudinal Study (e.g., Xue and Meisels, 2004) in the US. UNICEF's report on the quality of childhood, for example, employed social research methods to illustrate the poor standing of the UK on a range of indicators of quality of life for children, including relative poverty and family breakdown (Adamson, 2013). Further UNICEF findings have highlighted how children in the UK feel caught in a 'materialist trap' and do not spend enough time with their families (Ipsos-MORI and Nairn, 2011). A recent carefully conducted report by the National Society for the Prevention of Cruelty to Children (NSPCC) analysed data from a range of national databases in order to track trends in the extent and reporting of child abuse by region across the UK, with a view to developing a consistent methodology for monitoring rates of abuse annually across the country (Harker et al., 2013). Data of this nature could provide vital information for health and social care policy. Thorough scientific evidence is also critical in testing the effectiveness of intervention programmes aimed at promoting child development, and in determining how resources should be channelled so that the best outcomes for children can be achieved. In that context, social science methodology has played a leading role in rigorously evaluating the outcomes of major child development initiatives, such as the US Head Start Programme, the UK's SureStart Programme, the UK's recent National Evaluation of the Nurse Family Partnership, and Multi-Systemic Therapy for At-Risk Teens. While many of these national projects are supported by research funding agencies, quite a number are also funded directly by government departments, a fact which in itself illustrates both the importance of social science for supporting governmental health and social care strategies, and also the importance, and fruitfulness, of close partnership between academic institutions and government.

While the emphasis in this chapter is on social science, we would want to emphasise that child development research is fundamentally interdisciplinary and the best research – past, current and future – involves the creative team-working of researchers from a broad range of disciplines including psychologists, sociologists, biomedical scientists (neuroscience, physiology, genetics, pharmacology), epidemiologists, economists and statisticians. We strongly believe that the interdisciplinary character of child development research is vital for its continuing vigour as a field.

Mindful of the very incomplete picture we are able to paint of social science's contribution to this area, in this chapter we review two interrelated topics that we have been particularly involved in for some time: 1) parent–child attachment; 2) postnatal depression, and in each case we use the findings to elucidate broader conclusions regarding parenting and its influences on development. In doing so, we hope to show how social scientific thinking and research methods tackle questions of child development and where these fields are taking us in the future.

Parent–child attachment

Attachment theory is arguably the most influential account of the role played by the parent–child relationship in child development. Developed originally by the British Psychiatrist John Bowlby (Bowlby, 1969), it represents a unique integration of thinking from developmental psychology, evolutionary psychology, cognitive science and ethology. According to this model, a primary function (in the evolutionary sense) of a child's bond with his/her primary caregivers is to ensure the child's safety and protection against threats to his/her survival during the very protracted period of juvenile immaturity that characterises human development. Through a series of mechanisms that are still not completely understood, over the first year of life young infants develop stable and selective bonds to one or more consistent caregiving figures who they then selectively seek out in times of stress. During such times of stress or threat, the child's 'attachment system' is thought to be activated, which sets in motion a series of coordinated behaviours (supported by underlying physiology) whose function is to bring about proximity to a primary attachment figure. Once proximity is achieved, the child's sense of threat diminishes and the attachment system is deactivated, in what is conceptualised in an explicitly homeostatic fashion. Children enlist a rich, flexible and developmentally changing array of behaviours and communications in order to bring about proximity to a caregiver when stressed, from calling, crying and seeking in early life, to sophisticated communication and negotiation in later development.

A major impetus to the scientific study of attachment came from the work of the American psychologist Mary Ainsworth, who developed a pioneering structured observational tool for systematically measuring attachment behaviour in the laboratory, known as the Strange Situation Procedure (Ainsworth, Blehar, Waters, and Wall, 1978). This 21-minute procedure consists of a series of brief episodes in which natural cues to danger are presented to infants, including the appearance of a stranger and two brief periods when they are separated from their parent. These cues are assumed to activate the child's attachment system and provide opportunity for the researcher to observe how the child's attachment behaviour is expressed and particularly the way in which it appears to be organised to bring about proximity to, and gain comfort from, the attachment figure. Ainsworth's major discovery was that infants in the second year of life show marked and surprising variations in their attachment behaviour under these conditions. The majority, referred to as 'secure' or just 'B-type' behave as one would expect given the functions ascribed to attachment discussed above: when secure infants are separated they actively call and seek their caregiver, and upon reunion they quickly establish contact, which is effective in diminishing their distress and facilitating their return to exploration and play. A second category of infants, known as avoidant or type-A, show limited calling and seeking during separation, and actively avoid contact with the parent upon reunion, despite the fact that physiological markers suggest that they are equally aroused by the separation as other infants (Spangler and Grossmann, 1993; Sroufe and Waters, 1977; Zelenko et al., 2005). A third category, known as resistant or type-C, show marked distress when separated from a carer but upon reunion are unable to get comfort from their caregiver efficiently, either actively resisting contact when it is offered (e.g., angrily pushing away) or, less commonly, listlessly crying without seeking the parent's support.

These patterns of attachment behaviour are believed to represent adaptations on the part of the child to differences in the quality of parental care, and specifically in how responsive the caregiver is to the child's attachment signals, and how appropriate the parent's responses are. A fourth category was identified some time later by Mary Main and Judith Solomon (Main and Solomon, 1990), and is referred to as Disorganised or type-D. These infants appear unable to orchestrate a coherent way of dealing with the separation and reunion experience and show extended or momentary contradictory behaviours, such as seeking the parent and then strongly avoiding her, or freezing, stilling or rocking. These patterns of behaviour are thought to arise as a consequence of highly insensitive or frightening parental behaviour (Lyons-Ruth, Bronfman and Parsons, 1999; van IJzendoorn, Schuengel and Bakermans-Kranenburg, 1999).

Two key strands of attachment research have attempted to a) determine the nature of the environmental factors influencing the development of individual differences in attachment and b) charting the long-term consequences of these variations for children's socio-emotional development and liability to psychological disorder. In so doing, attachment research has gathered a host of evidence relevant to determining the potential of attachment as a productive target for intervention and prevention programmes. In the sections below we provide an inevitably selective summary of the findings of these studies and provide some examples of intervention studies that have arisen from research on attachment.

Environmental influences on attachment

A question of fundamental importance for the interpretation of the individual differences in attachment behaviour is whether they do indeed represent variations caused by environmental differences and not differences caused by the child's inherited genetic characteristics. Three key studies using samples of mono- and dizygotic twins have provided collectively compelling evidence that genetically-based variation in attachment behaviour in infants and toddlers is minimal (Bokhorst et al., 2003; O'Connor and Croft, 2001; Roisman and Fraley, 2008). For example, in one well-known study (Bokhorst et al., 2003), the estimate of heritability for attachment security in infants using the Strange Situation Procedure was zero, and all of the variance was attributable in roughly equal measure to the shared environment (environmental effects that make twins similar to each other) and the non-shared environment (environmental effects that make them different).

Furthermore, although some early small-scale studies suggested that insecure attachment, and disorganised attachment in particular, might be linked to certain gene polymorphisms (Lakatos et al., 2000), larger scale studies that focused on rigorous replication have failed to find any robust evidence of genetic association (Luijk et al., 2011). Collectively, a solid body of evidence thus testifies to one of the basic tenets of attachment theory, namely that variations in the organisation of attachment behaviour in early life are caused by the environment.

Since Ainsworth's pioneering early work, a host of studies have been conducted that have attempted to test the idea that the key environmental factor may be differences in the appropriateness and responsiveness of the parent to the child's attachment cues (referred to generally as the parent's 'sensitivity'). These studies, involving a wide range of procedures for directly observing mother–infant interaction, have been conducted in many countries around the world, and consistently uphold Ainsworth's original contention. Meta-analytic work has shown that the average effect size for these studies is equivalent to a correlation of $r = .21$ based on over 4000 independently sampled families (Cohen's $d = .43$), which is highly statistically significant (De Wolff and van IJzendoorn, 1997). This robust association is nevertheless moderate in size and although methodological factors undoubtedly play a substantial role in attenuating the observed association (such as the relatively brief observational periods often used when assessing sensitivity, and more generally the inherent measurement error in assessments of attachment and sensitivity), researchers have also taken this to imply that other factors, not well captured by the concept of behavioural sensitivity, must also play a part in the development of attachment. One particularly promising area of research has emerged in the last decade that is helping to shed further light on the processes that shape infants' attachments. Researchers are becoming increasingly interested in the way that parents *think* about their child, and particularly their capacity to imagine and make sense of their child's psychological states, such as their thoughts, feelings, motivations or focus of attention. For example, Slade and colleagues (2005) have developed an in-depth interview for parents regarding their relationship with their baby, which they carefully code for the degree and sophistication of mental state thinking that is apparent in their responses. The quality of the parent's capacities for reflectiveness, as measured in this way, has been found to predict later attachment security as well as the mother's parenting (Slade et al., 2005). Oppenheim, Koren-Karie and Sagi (2001), similarly, elicit parents' thoughts about their child by showing them clips of video recordings of them interacting with their child. Structured interview questions probe for parent's thinking about their baby's thoughts and feelings, which are again coded for the richness of their insights about their baby. Two separate studies from this group have also found this measure of maternal insightfulness to predict the infant's later attachment security (Koren-Karie, Oppenheim, Dolev, Sher and Etzion-Carasso, 2002; Oppenheim et al., 2001). Meins and colleagues have conducted perhaps the most comprehensive series of studies on the role of parental 'mind-mindedness' and attachment, by coding any spontaneous mind-related statements that parents make during their interactions with their infants. Four separate studies (Laranjo, Bernier and Meins, 2008; Lundy, 2003; Meins et al., 2012; Meins, Fernyhough, Fradley and Tuckey, 2001) have found that the tendency to treat the child as an individual with a mind, as reflected in the use of appropriate mind-related comments during interactions, is predictive of attachment security. In all these studies, the typical effects sizes are consistently and substantially higher than the meta-analytic average mentioned previously. Through detailed observational and interview-based studies, using carefully constructed and reliable coded methods, studies in this area have thus helped reveal some of the key features of the care-giving environment that appear to promote the development of secure attachment relationships and suggest a set of well-defined targets for preventive interventions. Before looking at attachment-based interventions, we turn first to a consideration of another critical prerequisite in the logic of prevention, namely whether or not there are long-term developmental benefits associated with secure attachment.

The long-term consequences of security and insecurity

A great deal of research has examined the developmental correlates of early secure and insecure attachment using longitudinal observational studies. Sroufe and Egeland (Erickson, Sroufe and Egeland, 1985) conducted what was arguably the seminal study of its kind in this area. This now-classic study focused on a sample of first-time mothers who were living in deprived circumstances (low socio-economic status, poor social support, multiple life stressors) and conducted detailed prospective assessments of the child's development, the parent–child relationship and the social-contextual factors impinging on the family. Among the wealth of important findings that emerged from this study, a critical observation was that children who showed insecure patterns of attachment in the Strange Situation at 12 and 18 months showed more behavioural problems at preschool as rated by their teachers relative to their secure counterparts, particularly problems related to hostility and impulsivity (Erickson et al., 1985). This pattern of greater behavioural problems among previously insecure children continued when the sample was followed up at school age, particularly in boys (Renken, Egeland, Marvinney, Mangelsdorf et al., 1989). Disorganised attachment in infancy was also found to be related to children's later emotional and behavioural problems in this sample, and in several others, including other high-risk samples. Carlson (Carlson, 1998) examined the long-term picture of overall adjustment and mental health among children in the Minnesota study, with a particular focus on disorganised attachment, and found that those who had been disorganised as infants had more internalising problems (anxiety and depression) in high school, more general symptoms of psychopathology as measured in a clinical interview at age 17 and more dissociative symptoms at age 19. Disorganised attachment in turn was related to the presence of a range of adverse social stressors in early life including single parenthood, poor parenting, child abuse and neglect. Similarly, Lyons-Ruth (1996) conducted a prospective study of deprived US infants and mothers and found that those who had been classified as disorganised in infancy had poorer cognitive development at 18 months and substantially elevated levels of externalising problems at age 7. Despite these individually quite compelling studies, it is not the case that every study that has followed up children whose attachment patterns were measured in early life have found that they were predictive of children's later outcomes. Fearon and colleagues (Fearon, Bakermans-Kranenburg, van IJzendoorn, Lapsley and Roisman, 2010) conducted a comprehensive meta-analytic review of all studies conducted to date in relation to children's externalising outcomes. Based on an analysis of 69 samples amounting to 5947 children, they found the overall association between attachment insecurity and later externalising problems to be moderately strong ($d = .31$) and highly significant. Importantly, this analysis showed that although each insecure category tended to show more externalising problems than secure children, the association was substantially higher for those who had been disorganised. The association was also stronger among boys, in clinical samples and when the outcome assessment used direct observation of the child's behaviour. Intriguingly, the connection between attachment and outcome did not diminish with time (i.e., when the outcome was assessed at later ages), and in fact showed signs of becoming stronger as children got older. This pattern of increasing behavioural problems with age was also found in a large US prospective study in which teacher reports were taken annually across the primary school years and related back to infant attachment (Fearon and Belsky, 2011).

In this study, disorganised boys from low-SES circumstances showed the most behavioural problems, particularly after the early primary school years. A further recent meta-analysis, this time based on 42 samples and 4614 children, showed that the connection between insecure attachment and internalising problems is rather weaker, and that avoidant, rather than disorganised, children are the group most affected. A final meta-analysis that has just been completed, summarising the results of 80 studies, has found that insecure attachment is quite robustly linked to children's later social competence, and indeed this association was the largest of the three examined in this set of meta-analyses (d = .39, Groh et al., under review). These findings are readily interpretable within the framework of attachment theory, because early attachment experiences are believed to form the basis of 'working models' of relationships that then guide a child's expectations, emotions and behaviour in later social interactions. The apparent prominence of social competence, and of aggression, would seem to suggest that, in childhood at least, peer relationships may be the most important social context within which these working models of attachment have their effects felt.

Attachment-based interventions

Over the last four decades attachment theory and research has generated a rich set of findings that fit coherently within a theoretical framework and collectively create a sound basis for intervention and prevention work, in which the enhancement of the security of the attachment relationship between a child and his/her caregivers is the therapeutic target. The majority of prevention studies that have been attempted so far have particularly focused on the construct of sensitivity and have sought to help parents increase their sensitive responsiveness to their infant's attachment cues. Existing interventions vary quite widely in how they do this and an exhaustive review is beyond the scope of this chapter (van IJzendoorn, Bakermans-Kranenburg and Juffer, 2003). Van den Boom's (1990) intervention represents a good example of one category of attachment intervention that is brief and focused on increasing maternal responsiveness to attachment cues. Infants who were highly irritable as neonates (prone to excessive crying and negative emotion) were randomised to either an observation-only control group or to the intervention. The intervention consisted of three home visits in which the mother was supported to: observe her infant's behaviour, following the infant's lead; and increase the contingency, consistency and appropriateness of her responses to her child's positive and negative cues. Following the intervention, 68% of the infants in the control group were classified as insecure, while only 28% were in the intervention group. Follow-up studies also suggested that the intervention was effective in improving wider developmental outcomes, including peer-cooperation at 42 months. The findings also indicated that the intervention may have enhanced *paternal* parenting, despite this not having been the direct focus of the programme. Not all interventions of this kind have been as successful though, and a meta-analytic review by Bakermans-Kranenburg and colleagues (2003) found that the average effect of the full array of interventions (88 different studies) was clearly significant and moderate in size (d = .20). Interventions that were brief, focused on sensitivity, and successful in enhancing sensitivity were the most successful in improving rates of attachment security. It is important to note though that several intensive and long-term interventions have also produced very promising results in high-risk contexts. For example, Cicchetti and colleagues (2006) and Moss and colleagues (Moss et al., 2011) have both reported very positive effects on disorganised attachment among young children who had been maltreated, using relatively intensive home visiting interventions (Cicchetti more intensive than Moss).

In summary, integrative developmental and clinical studies have produced a sizeable body of research that collectively paints quite a clear picture of whether and how one can support families effectively in order to promote the security of attachment. More work is needed to hone our understanding of the key factors that discriminate successful and less successful interventions, and to establish the degree to which early intervention of this nature produces long-term benefits in children's social and emotional development.

Postnatal depression

Overview

Postnatal depression (PND) affects about 14% of women in developed world populations; in developing countries, however, the rate is considerably higher – for example, in an impoverished community in South Africa, it was 35% (Cooper et al., 2009). Episodes of depression in the postnatal months show all the same features as depression occurring at other times, that is, pervasive low mood and loss of interest in normally enjoyable activities, as well as symptoms such as changes in sleep and appetite, and extreme feelings of guilt. In general, episodes of postnatal depression lift by six-to-nine months after childbirth, although a substantial minority of mothers can remain depressed right through the postnatal year and beyond, and those who experience postnatal episodes are at higher than average risk for the occurrence of further depression throughout their child's development (Cooper and Murray, 1995). The main risk factors for postnatal depression are the experience of depression and anxiety during pregnancy, and a lack of supportive relationships (particularly with a partner or the mother's own mother), as well as adverse living conditions. As well as the acute suffering and distress felt by women experiencing depression at such a vulnerable and significant time, research also indicates that post-natal depression can, in some cases, have adverse consequences for the children of affected mothers, and these effects can be felt through infancy into adulthood. This is particularly likely when the depression is severe and chronic.

A compelling body of research has investigated the impact of post-natal depression, which is shedding light on the range of areas of child development affected and the mechanisms that convey risk across development. Perhaps most importantly, this body of research has helped in the development and evaluation of clinical interventions aimed at supporting mothers experiencing PND, with the ultimate aim of reducing the impact that it has on their children's long-term development. The field of postnatal depression research represents an excellent example of how social-developmental research can help us understand child maladjustment, promote new innovative clinical practice and shape public policy. In addition, the insights gained from studying this particular population are valuable to revealing how parenting processes generally might influence child development, and what therefore needs to be targeted when seeking to prevent adverse child outcomes in diverse domains of functioning.

Postnatal depression and parenting

In the 1980s, groundbreaking work on PND was undertaken by a research group led by Field, Cohn, and Tronick who worked with women who were living in socioeconomically deprived conditions (Cohn, Matias, Tronick, Connell and Lyons-Ruth, 1986; Field, 1989). Detailed observational assessments were carried out in order to understand whether and how postnatal depression affected the pattern and quality of parenting of young infants. Mothers with postnatal depression were, on average, found to be more insensitive in their interactions than non-depressed controls, and were less able to respond sensitively to their infant's signals in a way that helped maintain the child's attention and regulate the child's emotions. The ways in which this insensitivity showed itself varied along two key dimensions: some depressed mothers would parent in an intrusive and irritable manner – abruptly interrupting the flow of the child's activities and appearing negative and hostile, whereas others would appear disengaged, emotionally flat and withdrawn. Similarly, the infants even at this young age (4–6 months) themselves behaved differently: they showed more negative emotion and were less socially responsive to their mothers. Importantly, these negative responses in infants of depressed mothers then extended to interactions with other people (Field et al., 1988). In subsequent studies by this and other groups, and in line with Field and colleagues' original observations, these quite distinct patterns of interaction between mothers and babies have been consistently found in the context of postnatal depression, especially when observed in groups living in conditions of hardship. In populations where background risk is low, however, although depressed mothers still find it difficult to respond sensitively to their infants, and may miss their infants' more subtle communication cues, the marked patterns of intrusiveness and disengagement seen in high-risk groups are not so evident, and accordingly, these infants are not as socially negative (Cohn, Campbell, Matias and Hopkins, 1990; Murray, Fiori-Cowley, Hooper and Cooper, 1996).

Beyond the infant's first year, another way in which the impact of PND has been measured has been to examine its effects on attachment. The research findings here have been quite complex, but have an important bearing in suggesting how the impact of PND may be reduced. The available evidence drawn from a number of studies suggests that infants whose mothers experience postnatal depression are at greater risk of developing an insecure attachment than infants of non-depressed mothers (Martins and Gaffan, 2000). Importantly, however, it is not necessarily the occurrence of PND in itself that can disturb attachment patterns. Instead, what seems critical is whether or not the mother's depression is severe and long-term in nature (Campbell et al., 2004). What is more, a recent study has suggested that the children of depressed mothers who were nonetheless able to interact sensitively and responsively with their infants, were less likely to be insecurely attached (Campbell, et al., 2004). The findings therefore strongly suggest that depression in mothers may adversely impact on the child's security of attachment to the extent that the depressive experience interferes with the parent's capacity to be sensitive and responsive to the child. In light of what we know about the later effects of attachment insecurity, these findings point to one important way in which the negative effect of postnatal depression on children's socio-emotional adjustment can be understood.

PND and children's later development

A sizeable body of research has examined the longer-term effects of exposure to postnatal depression on children's functioning. The effects so far identified span cognitive, linguistic and academic outcomes, as well as social adjustment, self-esteem and liability to psychological disorder (Murray, Halligan and Cooper, 2010). In the cognitive domain, studies have indicated that children exposed to postnatal depression obtain lower IQ scores in late infancy and preschool. This effect appears largely confined to those living in more adverse conditions, and where affected mothers have limited educational attainment themselves, and they may also be more evident in boys. Interestingly, recent work has indicated that when one takes into account the timing of the mother's depression, depression occurring in the first three months of life may have a specific negative effect on boys' later academic attainment (Hay et al., 2001; Murray, Arteche, Fearon, Halligan, Croudace and Cooper, 2010). Our long-term follow up study of the offspring of mothers with postnatal depression indicated that exposure to maternal depression can adversely affect formal exam performance at age 16 (Murray et al., 2010). Boys in our PND group scored on average 7 points lower (each lower point representing one lower GCSE grade) than the non-PND exposed controls, and this trajectory of poorer cognitive performance was evident when the children were as young as 18 months. Thus, there is quite consistent evidence that postnatal depression, particularly when it is severe and chronic, can negatively impact on children's cognitive and academic development. Notably, several longitudinal studies have traced these poor outcomes to the lower sensitivity and responsiveness of parental interactions with the infant and young child associated with PND (Milgrom, Westley and Gemmill, 2004; Murray, Kempton, Woolgar and Hooper, 1993). Apart from the negative effect on child cognitive functioning of this general reduction in responsiveness, studies have also found that depression may interfere with the 'attention grabbing and maintaining' features of the specialised speech that adults normally use with infants and young children (e.g., see Kaplan, Bachorowski and Zarlengo-Strouse, 1999), and that seems to be an important component of promoting infant learning. Thus, the speech of mothers who are depressed shows a preponderance of falling intonation contours, rather than the more musical and varied intonations of controls (Murray, Marwick and Arteche, 2010). Finally, depressed parents have been found to be less likely to engage in specific practices like book-sharing that are particularly beneficial to children's cognitive and language development (Bigatti, Cronan and Anaya, 2001). As for the attachment difficulties that are more common in infants of depressed mothers, this body of research on cognitive outcomes indicates, therefore, a set of particular aspects of parenting associated with the maternal disorder that may be fruitful to target in interventions if good cognitive development in the child is to be fostered.

Longitudinal studies also indicate with relative consistency that postnatal depression forecasts later behavioural and emotional problems in the child. For example, Murray (1992) found that infants of mothers with PND showed more behavioural problems, such as temper tantrums, separation anxiety and sleep problems, than controls at 18 months, despite maternal depression having remitted by this age in the great majority of cases. Several studies have documented similarly raised levels of behavioural problems in older children who were exposed to postnatal depression (e.g., Ghodsian, Zajicek and Wolkind, 1984; Sinclair and Murray, 1998), particularly in the presence of other related chronic stressors (e.g., subsequent depression, marital difficulties, paternal mental health problems, see Caplan, 1989; Cicchetti, Rogosch and Toth, 1998; Brennan et al., 2000; Hay et al., 2003; Hipwell, Murray, Ducournau and Stein, 2005). Our longitudinal study showed that here, too, quite specific features of parenting associated with postnatal depression were important in the development of such child problems – in this case, the hostility expressed towards the infant that was more common in depressed mothers in the postnatal months (hostility that was increased in the presence of marital conflict), provoked emotionally dysregulated behaviour in the infant that began to become a more general feature of their behaviour, leading to further vicious cycles of even more coercive and hostile parenting, and eventually conduct problems in the early school years (Morrell and Murray, 2003). In this domain, then, rather different difficulties in the parenting of depressed mothers are implicated in the development of behaviour problems, such as conduct disorder, from those associated with poorer cognitive functioning in the child.

Long-term follow-up studies in adolescence and adulthood also suggest that the offspring of mothers with postnatal depression are at increased risk of internalising problems. In a large community cohort study, Hammen and Brennan (2003) for example, found that the 15-year-old offspring of mothers with depression were more likely to have experienced a mental health problem, particularly depression, than non-exposed teenagers. As one might expect, this outcome was more likely when the maternal depression had been severe or of long duration. Recent findings from our own study (Murray, Arteche, Fearon, Halligan, Goodyer and Cooper, 2011) confirmed this association and suggest that multiple pathways may lead from postnatal depression to later child depression, including early insecure attachment and childhood emotional vulnerability, marital maladjustment, subsequent maternal depression and later insensitivity of care. In our study, each of these pathways appeared to add to the risk that children of mothers with PND would experience depression by age 16.

Together, the longitudinal and experimental studies show wide variability in how depression impacts on a mother's ability to respond to her infant and young child; correspondingly, the children of postnatally depressed mothers themselves are at increased risk for developmental difficulties across a wide range of outcomes. Importantly, those studies that have made direct observations of parenting and child development have begun to identify some of the parenting mechanisms mediating the associations between the maternal depression and adverse child outcome.

Treatment

In light of the significant distress and impairment experienced by mothers with depression and the significant negative consequences associated with it for their children, identifying mothers with PND and providing effective treatment are crucial objectives for healthcare policy. In recent years, public health policy has responded to the research evidence showing the importance of PND for parenting and child development, and in the UK, NICE guidelines set out procedures for screening all postnatal women for depression in the early weeks after childbirth. With regard to treatment, two recent meta-analytic summaries of existing treatment studies (Dennis and Hodnett, 2007; and Cuijpers, Brännmark and van Straten, 2008) found psychotherapeutic interventions (cognitive behaviour therapy, interpersonal therapy or counselling) to be moderately effective in reducing symptoms of depression in affected mothers. Broadly speaking, it is likely that effective interventions hasten the remission of postnatal depression, which shows high rates of remission even when untreated (Cooper et al., 1988). One critical limitation of the current evidence on treatment, given the relapsing nature of depression and the good evidence of its negative effect on child outcomes outside the infancy period, is the lack of long-term follow-up. It is also important to note that although there is some currently incomplete evidence that antidepressant medication may also be effective for postnatal depression, many mothers choose not to use it and hence its utility may be relatively limited (Appleby et al., 1997).

While there is good evidence then that psychotherapeutic treatments can improve symptoms of depression in mothers with postnatal depression, a vital question that has been addressed in a number of clinical studies is whether these interventions also have beneficial effects on child outcomes. The results of these studies have been quite mixed. One relatively large scale clinical trial for postnatal depression using randomised allocation to CBT, psychoanalytic therapy, counselling or a control group found positive effects on mother's depressive symptoms for all the active treatments, but limited evidence of significant effects on the child's subsequent development (Cooper et al., 2003; Murray, Cooper, Wilson and Romaniuk, 2003). Notably, similar findings were reported in two independent studies (Clarke, Tluczek, and Wenzel, 2003; and Forman et al., 2007), with both finding no evidence that the quality of mother–infant interaction improved following treatment, despite improvements in the mothers' depression. The evidence to date suggests therefore that it is necessary to directly target parenting in order to improve mother–infant interaction in the context of postnatal depression and associated difficulties (while also treating depression in the usual way). One large scale randomised controlled trial conducted in an impoverished community in South Africa, where risk for depression was high, delivered home visiting to mothers that combined a counselling approach with support for parents' social interactions with their infants, and in particular the management of distressed infant behaviour, and found that not only did depressive symptoms show some improvement, but infant attachment to the mother was more likely to be secure (Cooper et al., 2009). Despite these positive findings, few studies have shown, simultaneously, a positive intervention effect on the quality of mother–infant interaction and on long-term child outcomes, or indeed tested whether changes in parenting mediate treatment effects on child outcomes. These are critical issues to be addressed by future research. Given the evidence that those experiencing PND are more vulnerable to subsequent episodes of depression, and these too carry risks for adverse child outcomes, it would also seem important for mothers who experience the disorder postnatally to receive longer-term monitoring so that they could more easily receive further treatment should they relapse.

Conclusions and future directions

It is clear from the extensive research on children's attachment relationships and the effects of PND on child development that parenting has profound, long-lasting and wide-ranging effects. In this chapter, we focused on these two important examples to illustrate how the social science of child development can be used to tackle crucial questions regarding children's long-term social, cognitive and emotional development. In concluding, we would also want to draw attention to the fact that research shows compellingly that parenting – broadly conceived – plays a vital role in children's healthy social and emotional development generally, that is, beyond the realms of attachment and postnatal depression. Warm, supportive, consistent and authoritative parenting, sometimes collectively referred to as 'positive parenting' in the literature, confers a host of advantages for children right across childhood and adolescence. Furthermore, there is very strong evidence that parenting interventions for clinically referred children, particularly those based on social learning and cognitive-behavioural principles, can produce significant and substantial positive changes in both parenting and child outcomes, and can be done so in a cost-effective manner (Dretzke et al., 2005; Scott, 2010). A second important point to make is that while much of the existing research has focused on quite broad definitions of parenting (such as sensitivity or 'positive parenting') there is growing evidence that there is some specificity in the mechanisms driving child development. For example, specific aspects of parenting sensitivity, such as the quality of infant-directed speech, contingent responsiveness, and joint attention appear to be specifically linked to children's expressive and receptive language. Similarly, angry/hostile behaviour appears to be a particularly important factor in children's externalising problems, while anxious or overprotective care is associated with internalising problems. As the field progresses, we anticipate that an increasingly differentiated approach to interventions will emerge, which takes into account the individual profile of parenting problems and tailors therapeutic support to those processes in order to maximise their benefits and achieve better outcomes for children and families.

Bibliography

Adamson, P. (2013). *Child Wellbeing in Rich Countries: A Comparative Overview.* Florence: UNICEF.

Ainsworth, M. S., Blehar, M. C., Waters, E. and Wall, S. (1978). *Patterns of Attachment: A Psychological Study of the Strange Situation* (Vol. xviii). Hillsdale, NJ: Lawrence Erlbaum.

Appleby, L., Warner, R., Whitton, A., and Faragher, B. (1997). A controlled study of fluoxetine and cognitive-behavioural counselling in the treatment of postnatal depression. *British Medical Journal*, 314, 932–936.

Bakermans-Kranenburg, M. J., van IJzendoorn, M. H., and Juffer, F. (2003). Less is more: Meta-analyses of sensitivity and attachment interventions in early childhood. *Psychological Bulletin*, 129(2), 195–215.

Bigatti, S. M., Cronan, T. A. and Anaya, A. (2001). The effects of maternal depression on the efficacy of a literacy intervention program. *Child Psychiatry and Human Development*, 32, 147–162.

Bokhorst, C. L., Bakermans-Kranenburg, M. J., Fearon, R. M. P., van, I. M. H., Fonagy, P. and Schuengel, C. (2003). The importance of shared environment in mother–infant attachment security: a behavioral genetic study. *Child Development*, 74(6), 1769–1782.

Bowlby, J. (1969). *Attachment and Loss, Vol. 1: Attachment*. London: Hogarth Press and the Institute of Psycho-Analysis.

Bradshaw, J., Keung, A., Rees, G. and Goswami, H. (2011). Children's subjective well-being: international comparative perspectives. *Children and Youth Services Review*, 33(4), 548–556.

Brennan, P. A., Hammen, C., Andersen, M. J., Bor, W., Najman, J. M., Williams, and G. M. (2000). Chronicity, severity, and timing of maternal depressive symptoms: relationships with child outcomes at age 5. *Developmental Psychology*, 36, 759–766.

Campbell, S. B., Brownell, C. A., Hungerford, A., Spieker, S. J., Mohan, R. and Blessing, J. S. (2004). The course of maternal depressive symptoms and maternal sensitivity as predictors of attachment security at 36 months. *Development and Psychopathology*, 16(2), 231–252.

Caplan, H. L., Cogill, S. R., Alexandra, H., Robson, K. M., Katz, R. and Kumar, R. (1989). Maternal depression and the emotional development of the child. *British Journal of Psychiatry*, 154(June), 818–822.

Carlson, E. A. (1998). A prospective longitudinal study of attachment disorganization/disorientation. *Child Development*, 69(4), 1107–1128.

Cicchetti, D., Rogosch, F. A., and Toth, S. (1998). Maternal depressive disorder and contextual risk: contributions to the development of attachment insecurity and behavior problems in toddlerhood. *Development and Psychopathology*, 10, 283–300.

Cicchetti, D., Rogosch, F. A. and Toth, S. L. (2006). Fostering secure attachment in infants in maltreating families through preventive interventions. *Development and Psychopathology*, 18(3), 623.

Clark, R., Tluczek, A. and Wenzel, A. (2003). Psychotherapy for postpartum depression: A preliminary report. *American Journal of Orthopsychiatry*, 73, 441–454.

Cohn, J. F., Campbell, S. B., Matias, R. and Hopkins, J. (1990). Face-to-face interactions of postpartum depressed and nondepressed mother–infant pairs at 2 months. *Developmental Psychology*, 26, 15–23.

Cohn, J. F., Matias, R., Tronick, E. Z., Connell, D. and Lyons-Ruth, K. (1986). Face-to-face interactions of depressed mothers and their infants. In E. Z. Tronick and T. Field (eds), *Maternal Depression and Infant Disturbance* (pp. 31–45). San Francisco: Jossey-Bass.

Cooper, P. J. and Murray, L. (1995). Course and recurrence of postnatal depression. Evidence for the specificity of the diagnostic concept. *The British Journal of Psychiatry*, 166(2), 191–195.

Cooper, P J., Murray, L., Wilson, A. and Romaniuk, H. (2003) Controlled trial of the short- and long-term effect of psychological treatment of post-partum depression. I. Impact on maternal mood. *British Journal of Psychiatry*, 182, 412–419.

Cooper, P. J., Campbell, E. A., Day, A., Kennerley, H. and Bond, A. (1988). Non-psychotic psychiatric disorder after childbirth. a prospective study of prevalence, incidence, course and nature. *The British Journal of Psychiatry*, 152(6), 799–806.

Cooper, P. J., Tomlinson, M., Swartz, L., Landman, M., Molteno, C., Stein, A., et al. (2009). Improving quality of mother–infant relationship and infant attachment in socioeconomically deprived community in South Africa: randomised controlled trial. *British Medical Journal*, 338, b974.

Cronan, T. A., and Anaya, A. (2001). The effects of maternal depression on the efficacy of a literacy intervention program. *Child Psychiatry and Human Development*, 32, 147–162.

Cuijpers, P., Brännmark, J. G. and van Straten, A. (2008). Psychological treatment of postpartum depression: a meta-analysis. *Journal of Clinical Psychology*, 64, 103–118.

Dennis, C. L. and Hodnett, E. (2007). Psychosocial and psychological interventions for treating postpartum depression. *Cochrane Database of Systematic Reviews*.

De Wolff, M. and van IJzendoorn, M. H. (1997). Sensitivity and attachment: a meta-analysis on parental antecedents of infant attachment. *Child Development*, 68(4), 571–591.

Dretzke, J., Frew, E., Davenport, C., Barlow, J., Stewart-Brown, S. L., Sandercock, J., et al. (2005). The effectiveness and cost-effectiveness of parent training/education programmes for the treatment of conduct disorder, including oppositional defiant disorder, in children. *Health Technology Assessment*, 9(50), 1–250.

Erickson, M. F., Sroufe, L. A. and Egeland, B. (1985). The relationship between quality of attachment and behavior problems in preschool in a high-risk sample. *Monographs of the Society for Research in Child Development*, 50(1–2), 147–166.

Fearon, R. M. P., Bakermans-Kranenburg, M. J., van IJzendoorn, M. H., Lapsley, A. M. and Roisman, G. I. (2010). The significance of insecure attachment and disorganization in the development of children's externalizing behavior: a meta-analytic study. *Child Development*, 81(2), 435–456.

Fearon, R. M. P. and Belsky, J. (2011). Infant-mother attachment and the growth of externalizing problems across the primary-school years. *Journal of Child Psychology and Psychiatry*, 52(7), 782–791.

Field, T. (1989). Maternal depression effects on infant interaction and attachment behavior. In D.Cicchetti (ed.), *Rochester Symposium on Developmental Psychopathology, Vol.1: The Emergence of a Discipline* (pp. 139–163). Hillsdale, NJ: Erlbaum.

Field, T., Healy, B., Goldstein, S., Perry, S., Bendell, D., Schanberg, S. et al. (1988). Infants of depressed mothers show 'depressed' behavior even with nondepressed adults. *Child Development*, 59, 1569–1579.

Forman, D. R., O'Hara, M. W., Stuart, S., Gorman, L. L., Larsen, K. E. and Coy, K. C. (2007). Effective treatment for postpartum depression is not sufficient to improve the developing mother–child relationship. *Development and Psychopathology*, 19, 585–602.

Ghodsian, M., Zajicek, E. and Wolkind, S. (1984). A longitudinal study of maternal depression and child behaviour problems. *Journal of Child Psychology and Psychiatry, and Allied Disciplines*, 25(1), 91–109.

Golding, J. (1990). Children of the nineties. A longitudinal study of pregnancy and childhood based on the population of Avon (ALSPAC). *West of England Medical Journal*, 105(3), 80.

Hammen, C., and Brennan, P.A. (2003) Severity, chronicity, and timing of maternal depression and risk for adolescent offspring diagnoses in a community sample. *Archives of General Psychiatry*, 60, 253–258.

Harker, L., Jutte, S., Murphy, T., Bentley, H., Miller, P. and Fitch, K. (2013). *How Safe Are Our Children?* London: NSPCC.

Hay, D. F., Pawlby, S., Angold, A., Harold, G. T., & Sharp, D. (2003). Pathways to violence in the children of mothers who were depressed postpartum. *Developmental Psychology*, 39(6), 1083–1094.

Hay, D. F., Pawlby, S., Sharp, D., Asten, P., Mills, A. and Kumar, R. (2001). Intellectual problems shown by 11-year-old children whose mothers had postnatal depression. *Journal of Child Psychology and Psychiatry*, 42(7), 871–889.

Heckman, J. J. (2008). Schools, skills, and synapses. *Economic Inquiry*, 46(3), 289–324.

Hipwell, A. E., Murray, L., Ducournau, P. and Stein, A. (2005). The effects of maternal depression and parental conflict on children's peer play. *Child: Care, Health and Development*, 31(1), 11–23.

Ipsos-MORI and Nairn, A. (2011). *Children's Well-being in the UK, Sweden and Spain: The Role of Inequality and Materialism*. London: Ipsos MORI.

Kaplan, P. S., Bachorowski, J.-A. and Zarlengo-Strouse, P. (1999). Child-directed speech produced by mothers with symptoms of depression fails to promote associative learning in 4-month-old infants. *Child Development*, 70(3), 560–570.

Koren-Karie, N., Oppenheim, D., Dolev, S., Sher, E. and Etzion-Carasso, A. (2002). Mothers' insightfulness regarding their infants' internal experience: relations with maternal sensitivity and infant attachment. *Developmental Psychology*, 38(4), 534.

Lakatos, K., Toth, I., Nemoda, Z., Ney, K., Sasvari-Szekely, M. and Gervai, J. (2000). Dopamine D4 receptor (DRD4) gene polymorphism is associated with attachment disorganization in infants. *Molecular Psychiatry*, 5, 633–637.

Laranjo, J., Bernier, A. and Meins, E. (2008). Associations between maternal mind-mindedness and infant attachment security: investigating the mediating role of maternal sensitivity. *Infant Behavior and Development*, 31(4), 688–695.

Luijk, M. P. C. M., Roisman, G. I., Haltigan, J. D., Tiemeier, H., Booth-LaForce, C., van IJzendoorn, M. H., et al. (2011). Dopaminergic, serotonergic, and oxytonergic candidate genes associated with infant attachment security and disorganization? In search of main and interaction effects. *Journal of Child Psychology and Psychiatry*, 52(12), 1295–1307.

Lundy, B. L. (2003). Father- and mother-infant face-to-face interactions: differences in mind-related comments and infant attachment? *Infant Behavior and Development*, 26(2), 200–212.

Lyons-Ruth, K., Bronfman, E. and Parsons, E. (1999). Maternal frightened, frightening, or atypical behavior and disorganized infant attachment patterns. *Monographs of the Society for Research in Child Development*, 64(3), 67–96.

Lyons-Ruth, K. (1996). Attachment relationships among children with aggressive behavior problems: the role of disorganized early attachment patterns. *Journal of Consulting and Clinical Psychology*, 64, 32–40.

Main, M. and Solomon, J. (1990). Procedures for identifying infants as disorganized/disoriented during the Ainsworth Strange Situation. In M. T. Greenberg and D. Cicchetti (eds), *Attachment in the Preschool Years: Theory, Research, and Intervention. The John D. and Catherine T. MacArthur Foundation Series on Mental Health and Development* (Vol. xix, pp. 121–160). Chicago, IL, USA: University of Chicago Press.

Martins, C. and Gaffan, E. A. (2000). Effects of early maternal depression on patterns of infant–mother attachment: A meta-analytic investigation. *Journal of Child Psychology and Psychiatry and Allied Disciplines*, 41(6), 737–746.

Meins, E., Fernyhough, C., de Rosnay, M., Arnott, B., Leekam, S. R. and Turner, M. (2012). Mind-mindedness as a multidimensional construct: appropriate and nonattuned mind-related comments independently predict infant–mother attachment in a socially diverse sample. *Infancy*, 17(4), 393–415.

Meins, E., Fernyhough, C., Fradley, E. and Tuckey, M. (2001). Rethinking maternal sensitivity: Mothers' comments on infants' mental processes predict security of attachment at 12 months. *Journal of Child Psychology and Psychiatry*, 42(5), 637–648.

Milgrom, J., Westley, D. T. and Gemmill, A. W. (2004). The mediating role of maternal responsiveness in some longer term effects of postnatal depression on infant development. *Infant Behavior and Development*, 27(4), 443–454.

Morrell, J. and Murray, L. (2003). Parenting and the development of conduct disorder and hyperactive symptoms in childhood: a prospective longitudinal study from 2 months to 8 years. *The Journal of Child Psychology and Psychiatry and Allied Disciplines*, 44(4), 489–508.

Moss, E., Dubois-Comtois, K., Cyr, C., Tarabulsy, G. M., St-Laurent, D. and Bernier, A. (2011). Efficacy of a home-visiting intervention aimed at improving maternal sensitivity, child attachment, and behavioral outcomes for maltreated children: a randomized control trial. *Development and Psychopathology*, 23(1), 195.

Murray, L. (1992). The impact of postnatal depression on infant development. *Journal of Child Psychology and Psychiatry*, 33, 543–561.

Murray, L., Arteche, A., Fearon, R. M. P., Halligan, S., Croudace, T. and Cooper, P. (2010). The effects of maternal postnatal depression and child sex on academic performance at age 16 years: a developmental approach. *Journal of Child Psychology and Psychiatry*, 51(10), 1150–1159.

Murray, L., Arteche, A., Fearon, P., Halligan, S., Goodyer, I. and Cooper, P. (2011). Maternal postnatal depression and the development of depression in offspring up to 16 years of age. *Journal of the American Academy of Child and Adolescent Psychiatry*, 50, 460–470.

Murray, L. and Cooper, P. J. (1996). The impact of postpartum depression on child development. *International Review of Psychiatry*, 8(1), 55–63.

Murray, L., Cooper, P. J., Wilson, A. and Romaniuk, H. (2003). Controlled trial of the short- and long-term effect of psychological treatment of post-partum depression, 2. Impact on the mother–child relationship and child outcome. *British Journal of Psychiatry*, 182, 420–427.

Murray, L., Fiori-Cowley, A., Hooper, R. and Cooper, P. (1996). The impact of post-natal depression and associated adversity on early mother–infant interactions and later infant outcome. *Child Development*, 67, 2512–2526.

Murray, L., Halligan, S. L. and Cooper, P. J. (2010). Effects of postnatal depression on mother–infant interactions, and child development. In G. Bremner and T. Wachs (eds) *The Wiley-Blackwell Handbook of Infant Development* (pp. 192–220) Oxford, UK: Blackwell Publishing Ltd.

Murray, L., Kempton, C., Woolgar, M. and Hooper, R. (1993). Depressed mothers' speech to their infants and its relation to infant gender and cognitive development. *Journal of Child Psychology and Psychiatry*, 34(7), 1083–1101.

Murray, L., Marwick, H. and Arteche, A. (2010). Sadness in mothers' baby-talk predicts affective disorder in adolescent offspring. *Infant Behavior and Development*, 33(3), 361–364.

NICHD Early Childcare Research Network (2005). *Child Care and Child Development: Results From the NICHD Study of Early Child Care and Youth Development*. New York, NY: The Guildford Press.

O'Connor, T. G. and Croft, C. M. (2001). A twin study of attachment in preschool children. *Child Development*, 72(5), 1501–1511.

Oppenheim, D., Koren-Karie, N. and Sagi, A. (2001). Mothers' empathic understanding of their preschoolers' internal experience: Relations with early attachment. *International Journal of Behavioral Development*, 25(1), 16–26.

Renken, B., Egeland, B., Marvinney, D., Mangelsdorf, S. and Sroufe, L. A. (1989). Early childhood antecedents of aggression and passive-withdrawal in early elementary school. *Journal of Personality*, 57(2), 257–281.

Roisman, G. I. and Fraley, R. C. (2008). A behavior-genetic study of parenting quality, infant attachment security, and their covariation in a nationally representative sample. *Developmental Psychology*, 44(3), 831–839.

Sabates, R. and Dex, S. (2012). *Multiple Risk Factors in Young Children's Development.* London: Centre for Longitudinal Studies, Institute of Education.

Scott, S. (2010). National dissemination of effective parenting programmes to improve child outcomes. *The British Journal of Psychiatry*, 196(1), 1–3.

Sinclair, D. and Murray, L. (1998). Effects of postnatal depression on children's adjustment to school. Teacher's reports. *The British Journal of Psychiatry*, 172(1), 58–63.

Slade, A., Grienenberger, J., Bernbach, E., Levy, D. and Locker, A. (2005). Maternal reflective functioning, attachment, and the transmission gap: a preliminary study. *Attachment & Human Development*, 7(3), 283–298.

Spangler, G. and Grossmann, K. E. (1993). Biobehavioral organization in securely and insecurely attached infants. *Child Development*, 64(5), 1439–1450.

Sroufe, L. A. and Waters, E. (1977). Heart rate as a convergent measure in clinical and developmental research. *Merrill-Palmer Quarterly of Behavior and Development*, 23(1), 3–27.

van den Boom, D. (1990). Preventive intervention and the quality of mother–infant interaction and infant exploration in irritable infants. In Koops, W., Soppe, H., van der Linden, J. L., Molenaar, P. C. M., and Schroots, J. J. F. (eds), *Developmental Psychology Behind the Dikes* (pp. 249–270). Amsterdam: Eburon.

van IJzendoorn, M. H., Bakermans-Kranenburg, M. J. and Juffer, F. (2003). Less is more: Meta-analyses of sensitivity and attachment interventions in early childhood. *Psychological Bulletin*, 129(2), 195–215

van IJzendoorn, M. H., Schuengel, C. and Bakermans-Kranenburg, M. J. (1999). Disorganized attachment in early childhood: meta-analysis of precursors, concomitants and sequelae. *Development and Psychopathology*, 11, 225–249.

Walker, S. P., Wachs, T. D., Grantham-McGregor, S., Black, M. M., Nelson, C. A., Huffman, S. L., et al. (2011). Inequality in early childhood: risk and protective factors for early child development. *The Lancet*, 378(9799), 1325–1338.

Xue, Y. and Meisels, S. J. (2004). Early literacy instruction and learning in kindergarten: evidence from the early childhood longitudinal study-kindergarten class of 1998–1999. *American Educational Research Journal*, 41(1), 191–229.

Zelenko, M., Kraemer, H., Huffman, L., Gschwendt, M., Pageler, N. and Steiner, H. (2005). Heart rate correlates of attachment status in young mothers and their infants. *Journal of the American Academy of Child & Adolescent Psychiatry*, 44(5), 470–476.

2
Health and Wellbeing

James Campbell Quick, Robert J. Gatchel and Cary L. Cooper

Psychology and the social sciences play a central role in building a healthy world by bringing attention to supporting healthy families, encouraging healthy communities, and designing healthy workplaces (Rozensky et al., 2004). However, advancing health and wellbeing requires balanced attention to: chronic health risks and threats; the treatment of manifest health problems; and maximising (not optimising) health and wellbeing through positive psychological practices. The occupational context, or work environment context, is one especially important social context in which health and wellbeing can be effectively advanced (Macik-Frey et al., 2007). The behavioural and social sciences are so vitally important to the enhancement of health and wellbeing because so much of human behaviour is learned behaviour, not natural behaviour. That is, much human behaviour is socially constructed. Therefore, established learning principles offer a powerful and positive way to advancing health and wellbeing through the behavioural and social sciences. Three pathways for enhancing physical and mental health and wellbeing are:

- classical conditioning
- operant conditioning
- observational learning, or modelling.

In addition, this chapter addresses the environmental context, especially the work environment, as venues through which health and wellbeing can be enhanced. Consideration of negative environmental attributes, environmental risks, and bad relationships (especially, e.g., bad supervision in the workplace) are addressed. Social support IS central to health and wellbeing, in the workplace and beyond.

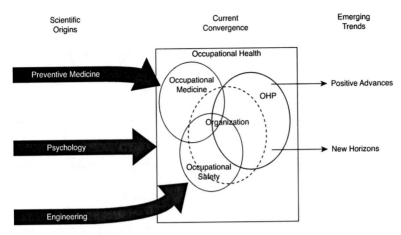

Figure 2.1 Occupational health in context
Source: M. Macik-Frey et al. (2007).

Drawing on the emerging field of positive organisational behaviour (Nelson and Cooper, 2007), positive psychology, and positive organisational scholarship, the chapter includes a section focused on positive psychological wellbeing. Positive psychological wellbeing is a strengths-based approach for advancing health and wellbeing through the behavioural and social sciences, while underpinning organisational wellbeing.

Preventive health management

Preventive health management is an encompassing extension of the philosophy, or theory, of preventive stress management™, which was originally conceived in the social context of the organisation (J. C. Quick and Quick, 1984; J. D. Quick et al., 1998). J. C. Quick (1999) broadened the fundamental philosophy, principles and practice to the wider range of chronic health problems beyond stress, such as violence, suicide, sexual assault and organisational injustice. This broader preventive health management framework has been specifically applied to chronic organisation problems like workplace violence (Mack et al., 1998) and sexual harassment (Bell et al., 2002). The advancement of positive health and wellbeing includes the prevention of negative wellbeing, the latter defined in terms of chronic health problems such as burnout, anxiety, depression and stress (Davidson et al., 2010). Enhancing health and wellbeing, and relieving the burden of suffering, in any population requires attention to both positive pathways and risks of negative or adverse outcomes (Macik-Frey et al., 2007). Figure 2.1 suggests how preventive medicine, psychology, and engineering can impact occupational health in context.

Principles and practices

Underlying the philosophy and theory are five principles that serve as guidelines for researchers and practitioners alike (see J. C. Quick et al., 2013). The first two principles set the tone for the practice of preventive health management, especially in occupational contexts and work environments.

Principle 1: Individual and Organisational Health are Interdependent
Principle 2: Leaders have a Responsibility for Individual and Organisational Health

These two principles reflect a philosophy that health and wellbeing, especially at work, can be accomplished through effective individual partnerships and collective social action. Health and wellbeing are therefore not seen as just attributes of individuals, but as the result of social and interpersonal processes in which individuals are engaged.

The power of preventive health management centres on its capacity to identify threats and risks to health and wellbeing; to identify asymptomatic and symptomatic health problems; and then serve as the basis for preventive intervention. The first function of the model is a surveillance function that enables professionals to identify health problems in social contexts. Chronic problems that threaten health and wellbeing have life histories that begin with health risks. Once these health risks and health problems are identified, the second function of the model is to bring to bear preventive interventions through the social sciences at one of three levels: primary prevention focuses on the health risk; secondary prevention on low intensity health disorders; and tertiary prevention (or treatment) on the more serious health problems (J. C. Quick et al., 2013).

From a public health and social science perspective, the preferred point of intervention is primary prevention, which prevents the onset of a health problem. Primary prevention is then supplemented with secondary prevention and, in turn, tertiary prevention when needed to help those who are suffering and experiencing frank health problems. Primary prevention methods focus on changing or managing the cause of the health risk or threat and, as such, are the most 'preventive' interventions. Primary prevention practices include environmental redesign and developing socially supportive relationships. Secondary prevention practices are designed to improve individual resilience and strength, and they are complementary to the primary prevention methods. Exercise, relaxation techniques and meditation are among the secondary prevention practices. Tertiary prevention can be considered professional treatment that is needed when the primary and secondary practices are not sufficient. At the tertiary level, the practices are aimed at healing the wounds and health problems for both individuals and organisations.

Biographical research application: three profiles emerge

In one extended application of the preventive health management model, we examined the health of leaders in business, other professions and politics, spanning the mid-1800s through to the outset of the twenty-first century (J. C. Quick et al., 2000). We gave special attention to strength factors such as character (Gavin et al., 2003), to health and wellbeing risks such as the absence of secure attachments (Joplin et al., 1990), and to individual vulnerabilities such as the inability to seek help when warranted (Joplin et al., 1995). Rather than focusing on environmental risks or individual exposures, we aimed to understand, from the 'inside out', the strengths and vulnerabilities or flaws in specific leaders, such as Theodore Roosevelt, Florence Nightingale, Lee Iacocca, Winston Churchill, Katherine Graham and Bill Gates (J. D. Quick et al., 2002). The outcomes from this method of inquiry yielded interesting results and three different profiles emerged in the research. These profiles or patterns were: health and wellbeing by building on strength; health and wellbeing through compensation (that is, conquering weakness); and health and wellbeing by living one's dream or following one's calling.

The first profile of health and wellbeing through strength is exemplified in the case of Bill Gates, whose focus and passion for computer software creation led him to build one of the largest businesses of his generation, Microsoft (J. D. Quick et al., 2002). Gates came to share with Andrew Carnegie a similar insight about wealth; that is, those who accumulate great wealth are only stewards of the money and owe back to the community from which they were enriched. In Gates' case, he and his wife Melinda have accepted the personal challenge of addressing World health. The Gates have donated well over $5.5 billion of their personal wealth to provide medical help and research for the underprivileged of the World. Thus, success through strength does not necessarily lead to self-centredness. Ebby Halliday is another distinguished American who exploited her gifts and abilities to build a real estate fortune and gave lavishly to those in need (Gavin et al., 2013).

The second profile of health and wellbeing through compensation is a different way that may be best exemplified in the case of Theodore Roosevelt (J. D. Quick et al., 2002). The story of his conquering child-hood asthma and physical limitations to become an adult known for his physical vigour, intellectual activity and moral conscience is part of American legend. Less well known may be the case of Norman Vincent Peale who became a renowned American preacher with a message on the power of positive thinking. He did so through the effortful activity of conquering his own negative thinking, doubts and uncertainties. The profile may require overcoming negative health and wellbeing.

The third profile of health and wellbeing through living your dreams or following your calling is a challenge for many successful women and men. Such was the case for Florence Nightingale in mid-nineteenth century Victorian England where a woman's place was to marry, have children and manage the home. From an early age, Nightingale over-came these social restrictions and limitations through great effort while overcoming in mid-life the health problems that plagued her until her death at age 90. She ultimately won the support of her father, her Queen and many more, while overcoming the guidance of her mother and some others who would have had her pursue a conventional path. The gifts she left to the World include the professionalisation of nursing, the advancement of public health and the incorporation of statistics and data in the study of health and illness. However, it is critical to acknowl-edge the significant negative health and wellbeing that Nightingale encountered in her years after the Crimean War, years in which she experienced significant physical and psychological suffering. Her nega-tive health and wellbeing did not ultimately, however, define her life. Nightingale is a true example of a life well lived in which the positive overcame the negative for good.

Strengths and limitations of biographical research

A real strength of biographical research is the consideration of the whole scope of individual lives to include attention to strengths, weaknesses and, importantly, adaptive or coping mechanisms engaged over the lifespan. Health and wellbeing are not short-term concerns, but rather long-term concerns with consequences often being time lagged. For example, Adolph Meyer of Johns Hopkins Medical School found, in the early 1900s, that major life change events, such as the death of a loved one or loss of employment, created health risk factors for individuals, but the manifest health problems often lagged by 12 to 18 months of the event. The behavioural and social sciences can be powerfully applied to create insights into health and wellbeing through biographies. On the other hand, a key limitation of biographical research is the difficulty to generalise and apply the findings. The challenge for the biographical researcher is to bring meaning and value to the work. However, there is a value in biographical research to potentially counterbalance the limitation. Biographical research can provide inspirational insights to the individual as they make life choices and key decisions for themselves. Herbert Simon's 1968 Karl Taylor Compton Lecture at MIT explored the sciences of the artificial. Rather than being instinctual, most of human behaviour is learned, acquired and socially constructed; hence, it is artificial. Therefore, the individual is left with the task and opportunity to create and live her or his own life. Learning through biography how others have lived their lives has value.

The use of learning principles to advance health and wellbeing

The behavioural and social sciences offer ways to advance health and wellbeing by drawing on well-established learning principles. We look first at three basic learning principles and then to their application.

Basic learning principles

In the context of health psychology, Baum et al. (1997) have reviewed the three major types of learning – classical conditioning, operant conditioning and observational learning (often called modelling). In terms of *classical conditioning*, the eminent Russian physiologist, Ivan Pavlov (1849–1936), first described this process with his work on the conditioned reflex. Reflexes are specific, automatic, unlearned reactions, produced by a specific stimulus. For example, if you ever touch a surface that you do not know is hot (such as a hot stove), you will most likely demonstrate a reflexive behaviour – the immediate withdrawal of your hand from the stove! Likewise, if a piece of dust suddenly gets into your eye, your eye will reflexively blink and secrete tears. These *unconditioned reflexes* occur automatically and have a great deal of survival value for you.

Pavlov set out to demonstrate that such unconditioned reflexes could be *conditioned*, producing what came to be known as *classical conditioning*, which is one of the most basic forms of learning. He did this by conducting a series of well-known studies on the process of classical conditioning using dogs as experimental subjects. In these studies, Pavlov evaluated situations in which a neutral stimulus or event (such as a bell) was presented to a dog just prior to the presentation of food (an unconditioned stimulus that normally elicits an automatic unconditioned reflex of salivation). After a number of such presentations, the bell (now a conditioned stimulus) would elicit a conditioned or learned salivation response when presented by itself in the absence of food. The conditioned reflex of salivation occurred to the bell alone. This represents the process of classical conditioning, and it is based on the learned association or connection between two stimuli, such that the bell is associated with food (with both having occurred together at approximately the same point in time). An association is learned between a weak stimulus (such as the bell) and a strong stimulus (such as the sight of food) so that the weak stimulus comes to elicit the response (i.e., salivation) originally controlled only by the stronger one.

The second type of learning – *operant or instrumental conditioning* – was originally formulated by Edward Thorndike, and then more thoroughly developed by B. F. Skinner. In contrast to classical conditioning, operant conditioning develops new behaviours that produce positive consequences or remove negative events. Thus, behaviours that produce food, social approval and other positive consequences (such as money for working), or that reduces damaging or aversive events (such as stopping at a red light in order to avoid receiving a traffic ticket), illustrate operant behaviours. These behaviours 'operate' on the environment in order to bring about changes in it. For centuries, operant conditioning methods were used for training circus animals. When an animal produced a desired behaviour (e.g., rolling over or jumping through a hoop), reinforcement in the form of food would be provided to the animal. Reinforcement refers to any consequence/event that increases the likelihood that a particular behaviour will be repeated (or that strengthens such a behaviour).

Finally, the third form of learning is *observational learning* (which is often also referred to as *modelling*). This form of learning refers to learning that occurs without any apparent direct reinforcement. Indeed, many behaviours, especially in children, are acquired simply by observing those behaviours performed or modelled by another person. The old adage 'Do as I say, and not as I do' completely runs counter to observational learning principles.

Examples of learning in real-life situations

As reviewed by Gatchel et al. (2009), these types of learning can be readily seen when trying to understand the behaviour of an individual who may be hurt at work. Consider the individual who may have hurt his/her back while lifting objects at work. This person may now, after telling co-workers about the pain, begin to display *pain behaviours,* such as verbal expressions of hurting (e.g., moaning, sighing, complaining, etc.), as well as nonverbal behaviours such as limping, grimacing, etc. If co-workers sympathise with the worker, and offer to perform any lifting when required, they may be inadvertently operantly conditioning these pain behaviours to the point where the worker allows his/her co-workers to assume all of his/her work requiring lifting. In the long run, this may not be a positive scenario if these pain behaviours persist because of the reinforcement for not lifting.

In terms of classical conditioning, this type of learning may also produce avoidance of lifting by a worker. At first, before classical conditioning had a chance to occur, there was no association between lifting objects at the job and the avoidance of lifting because of fear of pain. However, if the worker begins to experience back pain while lifting, which becomes progressively worse over time, then he/she may now hesitate to lift anything because of fear of potentially exacerbating the back pain which he/she is already experiencing. Thus, after this classical conditioning, any prompting or requirement to lift an object automatically produces a fear response and an active avoidance of any lifting in order to avoid hurting the back more. There is now a classically conditioned negative emotional response of lifting objects at work because of the fear of pain. Coupled with the previous discussed operant conditioning process that may also be happening, it is not difficult to imagine why these workers may refuse any lifting requirements at work. Finally, it may also be the case that this worker watched television which had an ad for a new analgesic on the market to help eliminate back pain due to strenuous activities, such as lifting. Here again, this observational learning will likely further prompt the avoidance of lifting objects at work.

Of course, the above are examples of learning principles that may negatively affect job-related lifting. However, one can 'reverse' such avoidance behaviour by changing the learning contingencies/reinforcements. For example, one can be somewhat sympathetic about a co-worker's 'bad back' but, at the same time, encourage him/her to lift lighter weight objects at first (i.e., reinforce positive lifting behaviour). Moreover, the worker now may begin to realise that lifting these lighter weight items no longer produces pain. This results in the 'breaking' of the classically conditioned association between lifting and fear of pain. With the renewed confidence, the worker may now also not be 'sensitised' by television ads about analgesics and back pain caused by lifting.

Thus, such learning principles exert an influence on our behaviour in everyday situations.

There are also other common examples of how these three types of learning can have a negative impact on healthy behaviours. For example, take the case of smoking in the workplace. Often, a new non-smoking worker may enter a work environment where many co-workers, who he/she begins to socialise with, frequently smoke. This may set the stage for modelling/observational learning, prompting the new worker to want to be 'part of the group' for social support purposes. As a result, he/she begins to also smoke when socialising/working with them. If these co-workers also communicate that they smoke because it reduces the stress and boredom of work, the new worker may also begin to smoke to see if this is true for him/her. If it does have this positive effect, then this worker develops a classically conditioned association between smoking and the reduction of job stress and boredom. Moreover, whenever he/she feels stressed or bored, this worker automatically 'pulls out' a cigarette in order to reduce these negative feelings (this is a case of operant behaviour). Thus, as can be seen, negative environmental factors (i.e., smoking in the workplace) can prompt the development of a high-risk health behaviour in a formerly non-smoking individual. This can also occur in other non-healthy behaviours, such as self-medication, eating 'junk food', etc.

Positive psychological wellbeing – positive organisational wellbeing

Positive wellbeing in the workplace has grown in popularity over recent years, partly as a consequence of the global recession and downturn. We have the former President of France, Sarkozy, highlighting the concept of Gross National Wellbeing, Prime Minister Cameron of the UK measuring annually the national wellbeing of the population, and we have the Bhutan declaration that the goal of countries should be to enhance their national wellbeing, signed by 79 countries at the UN in April of 2012. Within the workplace, there has been a movement to assess an organisation's hedonic and eudaimonic wellbeing, and their impact on organisational outcomes. Hedonic refers to subjective wellbeing of happiness and positive emotions (Diener, 2000), whilst eudaimonic refers to the purposeful aspects of wellbeing, which Ryff et al. (2004) broke down into roughly six aspects of self-acceptance, environmental mastery, positive relationships, personal growth, purpose in life and autonomy. There have been many ways of measuring these aspects with respect to individual wellbeing, the PsyCap Questionnaire being the most widely used, assessing self-efficacy, optimism, hope and resiliency.

These aspects of positive psychological wellbeing have been extended into the workplaces, but with different organisational measures like engagement, job control and autonomy, positive relations with others, etc., as well as organisational outcome measures that impact the bottom line such as productivity, sickness absence, etc. Harter et al. (2003), for example, did a large scale study of nearly 8000 business units in 36 companies exploring positive psychological wellbeing and a range of outcome measures like customer satisfaction, productivity, profitability, employee turnover and sickness absence. They found a strong relationship between employee wellbeing and productivity, as well as a link between positive wellbeing and engagement and job satisfaction. Attridge (2009) reviewed many of the studies linking positive psychological wellbeing and organisational benefits, categorising them in terms of outcomes such as lowered sickness absence, customer loyalty, productivity, 'better return for investors', increases in operating income, etc.

As a caveat, we should note here that the positive and the negative in organisational life are frequently intertwined or at least proximate. Therefore, it is important to acknowledge the negative because there is frequent pain, injury and damage done at work, though certainly not all or even most are intentional. Cameron (2007) makes this point eloquently and offers forgiveness as an antidote for the negative experiences in work life. He recognises that it can be difficult if not impossible to discuss and explore the positive aspects of organisational life without acknowledging the negative. With this acknowledgement, we choose to place our primary emphasis on the positive.

The ASSET model of workplace wellbeing

Robertson and Cooper (2011) created what they termed the ASSET model of workplace wellbeing. They identified the key workplace factors as: resources and communication; control/autonomy; work–life balance/workload; job security and change; work relationships; and job conditions. Organisational outcomes were moderated by the psychological wellbeing factors of sense of purpose and positive emotions. The organisational outcomes were measured by productivity/performance, attendance, retention, attractiveness of recruits and customer satisfaction, whilst the individual ones by job satisfaction/performance indicators, morale, good citizenship and health. Extensive research over the last ten years is now available to suggest that positive organisational wellbeing can lead to measureable outcomes in the productivity of individuals,

health of employees and bottom line indicators of economic performance in businesses (Dewe and Cooper, 2012).

As Senator Bobby Kennedy said in 1968, although the Gross National
Product (GNP) and Gross Domestic Product (GDP) is important in
society so are the other human factors:

> Too much and for too long, we seemed to have surrendered personal
> excellence and community values in the mere accumulation of mate-
> rial things. Our GNP is $800b a year, but that GNP, if we judge the
> USA by that, that GNP counts air pollution and cigarette advertising
> and the ambulances to clear our highways of carnage. It counts
> special locks for our doors and the jails for the people who break
> them. It counts the destruction of the redwood and the loss of our
> natural wonder in chaotic sprawl... Yet the GNP does not allow for
> the health of our children, the quality of their education or the joy of
> their play. It does not include the beauty of our poetry or the strength
> of our marriages, the intelligence of our public debate or the integrity
> of our public officials. It measures neither our wit nor our courage,
> neither our wisdom nor our learning, neither our compassion nor our
> devotion to our country, it measures everything in short, except that
> which makes life worthwhile.

Well-Being Index® and health enhancement

Since 2008, the Gallup-Healthways Well-Being Index® has aimed to
offer an index of national wellbeing in the US, much as the Dow Jones
Industrial Average has aimed to provide a benchmark of the nation's
economic health and wellbeing. The concept is both worthy and novel
(see http://well-beingindex.com). In the case of the Dow Jones, there
are a set of Blue Chip companies that compose the industrial average.
For the US Well-Being Index, there are a set of six basic dimensions that
compose the composite index. These are:

- Life evaluation
- Emotional health
- Physical health
- Health behaviours
- Work environment
- Basic access.

The basic access is to necessities crucial to high wellbeing. There are 13
items in the basic access category: community satisfaction, community

improvement, clean water, medicine, safe place to exercise, affordable fruits and vegetables, safety in walking alone at night, enough money for food, enough money for shelter, enough money for healthcare, visit to a dentist, access to a doctor, and access to health insurance. The life evaluation measure ranges from suffering (0–4) through struggling (5–6) to thriving (7–10). The US Well-Being Index is determined through a Gallup-Healthways surveying methodology. No fewer than 500 US adults nationwide are interviewed each day, with the exception of major holidays. For primarily Spanish-speaking respondents, interviews are conducted in Spanish. The daily samples are composed of over 850 landline respondents and 150 cell phone respondents. Samples are weighted by gender, age, race, Hispanic ethnicity, education region, adults in the household and phone factors.

Individuals and organisations alike can reap the positive benefits of health and wellbeing. Merrill, et al. (2013) found in a study of 20,000 American workers that an integrated approach to health and wellbeing improvement (a benefit to the individual) could help maximise employee job performance while reducing absenteeism (a benefit to employers). The researchers found that the benefits accrued from healthy eating and exercise. Three key findings related to job performance were:

- 25% greater likelihood of higher job performance for those who ate healthy for the entire day.
- 20% greater likelihood to have higher job performance for those who ate five or more servings of fruit and vegetables on four or more days in a week.
- 15% greater likelihood to have higher job performance for those who exercised for 30 or more minutes on three or more days in a week.

In addition, the researchers took an extensive look at obesity's impact on job performance and found the following:

- Job performance was 11% higher among employees who were not obese;
- Employees with well-managed chronic diseases were more highly productive than obese employees with chronic diseases who did not exercise;
- Obese employees, as well as those with a history of chronic disease and conditions related to pain and activity limitations, were more likely to be absent; and

- Obese employees had lower job performance and higher absenteeism than employees with depression and other chronic diseases or conditions.

One of the significant notes related to this research being reported through the occupational and environmental medicine community is the fact that the core of the intervention work is as much social science as it is medicine. Health and wellbeing at work are not just medical issues, but are ones where social science has much to contribute. We see that in the lifestyle and behavioural interventions in the research: eating behaviour, exercise and activity, weight management and reduction, and employee engagement strategies. These are all at the nexus of the social sciences intertwining with the medical sciences in order to produce a significantly better result for people.

There is a business case to be made for health and wellbeing in the workplace, and an economic case to be made for attending to health risks and hazards. Certainly, the social and behavioural sciences are attuned to the humanitarian case for health and wellbeing but the emerging evidence for the business case is increasingly strong.

Conclusions

The evidence has shown, in one major industrial restructuring case with over 13,000 personnel impacted, that the practice of preventive stress management™ can be lifesaving (J. C. Quick et al., 2013). We suggest at the close of this chapter that positive psychology and social science can be life-giving as they play a critical part in building a healthy world, especially in the workplace. Thus, advancing health and wellbeing requires preventing injury, harm and damage (i.e., negative wellbeing), while simultaneously maximising (not optimising) health and wellbeing through positive psychological practices. The work context is one especially important social context in which health and wellbeing can be effectively advanced because that is a common venue for the working people of the world.

The behavioural and social sciences are keys to health and wellbeing because most human behaviour is learned behaviour, in contrast to natural or instinctual behaviour. Because the majority of human behaviour is socially constructed, the reliance on learning principles is a positive way to health and wellbeing. Three specific pathways that we explored in the chapter were: classical conditioning, operant conditioning and observational learning, or modelling. Learning in the work environment context thus advances health and wellbeing.

To this learning framework we added new scientific findings from positive organisational behaviour (Nelson and Cooper, 2007), positive psychology and positive organisational scholarship (Cameron, 2007). In addition to the intuitively understood humanitarian case for health and wellbeing, we added the business case and economic evidence for organisational wellbeing as an extension of psychological wellbeing.

Bibliography

Attridge, M. (2009). Measuring and managing employee work engagement: a review. *Journal of Workplace Behavioral Health*, 24, 383–398.

Baum, A., Gatchel, R. J. and Krantz, D. S. (eds) (1997). *An Introduction to Health Psychology*, (3rd edn). New York: McGraw-Hill.

Bell, M. P., Quick, J. C. and Cycota, C. (2002). Assessment and prevention of sexual harassment: an applied guide to creating healthy organizations. *International Journal of Selection and Assessment*, 10(1/2), 160–167.

Cameron, K. S. (2007). Forgiveness in organizations. In D. L. Nelson and C. L. Cooper (eds), *Positive Organizational Behavior* (129–142). Thousand Oaks, CA: SAGE.

Davidson, O. B., Eden, D., Westman, M., Cohen-Charash, Y., Hammer, L. B., Kluger, A. N., Krausz, M., Maslach, C., O'Driscoll, M., Perrewé, P. L., Quick, J. C., Rosenblatt, Z. and Spector, P. (2010). Sabbatical leave: who gains and how much? *Journal of Applied Psychology*, 95(5), 953–964.

Dewe, P. and Cooper, C. L. (2012). *Wellbeing and Work*, Basingstoke: Palgrave Macmillan.

Diener, E. (2000). Subjective wellbeing. *American Psychologist*, 55, 34–43.

Gatchel, R. J., Kishino, N. D., Theodore, B. and Noe, C. (2009). Pain and learning. In J. C. Ballantyne, J. P. Rathmell and S. M. Fishman (eds), *Bonica's Management of Pain* (4th edn). Philadelphia, PA: Lippincott, Williams & Wilkins.

Gavin, J. H., Quick, J. C., Cooper, C. L. and Quick, J. D. (2003). A spirit of personal integrity: the role of character in executive health. *Organizational Dynamics*, 32, 165–179.

Gavin, J. H., Quick, J. C. and Gavin, D. J. (2013). *Live Your Dreams, Change The World*. Riverside, NY: American Mental Health Foundation.

Harter, J. K., Schmidt, F. L. and Keyes, C. L. M. (2003). Wellbeing in the workplace and its relationship to business outcomes. In C. L. M. Keyes and J. Haidt (eds), *Flourishing, Positive Psychology and Life Well-lived*, Washington, DC: American Psychological Association.

Joplin, J. R. W., Nelson, D. L. and Quick, J. C. (1990). Attachment behavior and health: relationships at work and home. *Journal of Organizational Behavior*, 20, 783–796.

Joplin, J. R., Quick, J. C., Nelson, D. L. and Turner, J. C. (1995). Interdependence and personal well-being in a training environment. In L. R. Murphy, J. J. Hurrell, Jr., S. L. Sauter and G. P. Keita (eds), *Job Stress Interventions* (309–321). Washington, DC: American Psychological Association.

Macik-Frey, M., Quick, J. C. and Nelson, D. L. (2007). Advances in occupational health: from a stressful beginning to a positive future. *Journal of Management*, 33, 809–840.

Mack, D. A., Shannon, C., Quick, J. D. and Quick, J. C. (1998). Chapter IV – Stress and the preventive management of workplace violence. In R. W. Griffin, A. O'Leary-Kelly and J. Collins (eds), *Dysfunctional Behavior in Organizations – Volume 1: Violent Behavior in Organizations* (119–141). Greenwich, CN: JAI Press.

Merrill, R. M., Aldana, S. G., Pope, J. E., Anderson, D. R., Coberley, C. R., Grossmeier, J. J., Whitmer, R. W. and HERO Research Study Subcommittee. (2013). Self-rated job performance and absenteeism according to employment engagement, health behaviors, and physical health. *Journal of Occupational & Environmental Medicine*, 55(1), 10–18.

Nelson D.L. & Cooper C.L. (eds) (2007). *Positive Organizational Behavior*. Thousand Oaks, CA: Sage.

Quick, J. C. (1999). Occupational health psychology: The convergence of health and clinical psychology with public health and preventive medicine in organizational contexts. *Professional Psychology: Research and Practice*, 30(2), 123–128.

Quick, J. C. and Quick, J. D. (1984). *Organizational Stress and Preventive Management*, New York: McGraw-Hill.

Quick, J. C., Gavin, J. H., Cooper, C. L. and Quick, J. D. (2000). Executive health: building strength, managing risks. *Academy of Management Executive*, 14, 34–44.

Quick, J. C., Wright, T. A., Adkins, J. A., Nelson, D. L. and Quick, J. D. (2013). *Preventive Stress Management in Organizations, Second Edition*, Washington, DC: American Psychological Association.

Quick, J. D., Cooper, C. L., Gavin, J. H. and Quick, J. D. (2002). Executive health: building self-reliance for challenging times. In C. L. Cooper and I. T. Robertson (eds), *International Review of Industrial and Organizational Psychology* (187–216). West Sussex, England: John Wiley & Sons, Inc.

Quick, J. D., Nelson, D. L. and Quick, J. C. (1998). The theory of preventive stress management in organizations. In C. L. Cooper (ed.), *Theories of Organizational Stress* (246–268). Oxford: Oxford University Press.

Robertson, I. and Cooper, C. L. (2011). *Wellbeing: Productivity and Happiness at Work*, Basingstoke: Palgrave Macmillan.

Rozensky, R. H., Johnson, N. G., Goodheart, C. D. and Hammond, R. (2004). *Psychology Builds a Healthy World*, Washington, DC: American Psychological Association.

Ryff, C. D., Singer, B. H. and Love, G. D. (2004). Positive health: connecting wellbeing with biology. *Philosophical Transactions of the Royal Society*, 359, 1383–1394.

3
Climate Change and Society

John Urry

The discovery of changing climates

Contemporary societies are faced by a new spectre haunting the 'globe' – the changing of the world's climate. This was not believed possible by scientists until fairly recently although the theoretical idea of a 'greenhouse effect' has been well established for a century or so.

That there are raised levels of 'greenhouse gases' in the atmosphere results from data collected from a measuring station established on the Mauna Loa volcano on Hawaii in 1959 to monitor CO_2 and other emissions in the atmosphere. These readings showed that CO_2 was not being fully absorbed into the oceans and was inexorably rising. This may be the most widely reprinted set of natural science data ever collected. This one almost accidental curve derived from data from one observatory in one somewhat obscure location showed how the earth was being irreversibly changed by unprecedented human activities that raised CO_2 levels.

Moreover, other data collected around the world recorded increasing temperatures on land and at sea. It was concluded that rising emissions were at least in part responsible for these increasing temperatures. And if temperatures continued to increase by anything between 2–6 degrees Celsius as emissions rise and stay in the atmosphere for hundreds of years, then human, animal and plant life will be transformed. The material world apparently does matter and can 'bite back' with interest (see leading US climate scientist, Hansen, 2011).

Two groups of analysts have dominated the understanding of apparently rising emissions and temperatures, namely physical scientists and economists (see Urry, 2011: chapter 1). On the one hand, there are many climate scientists working in and across a wide range of different scientific disciplines. Much of their work is synthesised every few years in the authoritative Intergovernmental Panel on Climate Change (IPCC) Reports. The IPCC was founded in 1988, a year of record temperatures, by the United Nations Environment Programme with up to 2500 scientists examining the links between greenhouse gas (GHG) emissions and climate change. These reports are endorsed by most governments, with the first published in 1990. By 2007 the IPCC stated that the evidence of humans changing climate is 'unequivocal'.

Pertaining to this, Nobel prize-winner Paul Crutzen argued that there is a new geological period of human history, the 'anthropocene' following on from the holocene. In this new period it is human activities that exert a major impact upon almost all aspects of the earth system, an impact equivalent to a great force of nature (available at http://www. anthropocene.info/en/anthropocene and accessed 18 September 2012). Drawing upon research with at least 47,000 peer-reviewed articles published by the mid-2000s are many scientific journalists who interpret and disseminate the sciences of climate change for a wider public (Pearce, 2007).

On the other hand, this scientific research shows that climate change is not a purely 'scientific' problem and that human actions are central to this apparent warming of the planet. Also research shows that such warming will only be slowed down or reduced if 'humans' around the world behave differently. Economists are typically viewed as being best able to examine these 'human' dimensions of global climate change. Their central role was especially reflected in the *Stern Review* which established the relatively limited costs of developing appropriate policies for mitigation by comparison with the vast expense of climates actually changing (Stern, 2007). Because 'economics' got in first, it monopolised the framing of humans in climate change understanding and debate. This led to a focus on human practices as individualistic, market-based and calculative, and thus generated responses to climate change based on individual calculation to change behaviour, new technologies to fix the problem and developing markets for novel 'green products'. Most major studies of climate change issues involve economic analysis.

It is important to note though that in the later sections of the monumental Stern Review, aspects of society and social customs enter analysis but are not thought to be fully explicable through the lens of 'economics' (Stern, 2007). There are some limits of economic models here. Thus even this Stern Review would suggest that a much wider range of social sciences needs to be brought to bear on these absolutely crucial issues.

So while economic institutions are globally significant this is often because of their social and political consequences and not just because of their role within markets. Large global corporations, such as those central to 'carbon capitalism', control the lives of workers and consumers and do not merely affect 'markets'. Many such corporations have huge vested interests in some version of 'business as usual' (see Urry, 2013, on carbon capital).

Moreover, there is a major problem in how economists treat energy. Economists typically regard energy as responsible for about 5% of the GDP of any economy because this is roughly what it costs. But we now know that fossil fuel-based energy is a unique bundle of commodities which are non-renewable and generate 'external diseconomies' on such a historical and geographical scale that they change climates and future supplies of energy, water and food. Some commentators calculate that if one of the three main fossil fuels, oil, is between ten and fourteen times more productive than the neo-classical model estimates, then the 'real' price of oil should also have been this much higher over the twentieth century, the 'century of oil' based on exceptionally cheap and available oil (Urry, 2013). Such a higher price would have led to a very different and much less mobile twentieth century.

Indeed, energy in general and oil in particular are not just any commodities. As Schumacher argues: 'There is no substitute for energy. The whole edifice of modern society is built upon it...it is not "just another commodity" but the precondition of all commodities, a basic factor equal with air, water, and earth' (quoted in Kirk, 1982: 1–2). Each year an astonishing eight billion tons of oil, gas and coal are used to 'energise' the systems of producing and consuming goods and services around the world. Modern lives totally depend upon burning these fossil fuels, to heat, power, manufacture and move people and objects.

A further problem in economic models is that most of the time people do not behave as individually rational economic consumers, maximising their utility from the basket of goods and services they purchase and use. People are creatures of social routine and habit. These routines stem from the many ways that people are locked into social practices and social institutions, including families, households, friendship groups, social classes, genders, work groups, businesses, schools, ethnicities, age cohorts, nations and so on. It is these institutions that organise and structure people's lives. Buying and using goods and services help to constitute these institutions and their typical social practices and it is such practices that are the very stuff of life. Shove thus argues against the restricted model of behavioural change based upon the dominant paradigm of 'ABC' – attitude, behaviour and choice, and argues in favour of transforming or replacing these very social practices so as to reduce energy 'demand' (Shove, 2010; Shove et al., 2012).

Significantly, though, changes do take place in production/consumption systems and they can occur very rapidly, such as the development of mobile communications. In a way this came from nowhere and transformed lives through new mobilised social fashions. So fixed and stable routines may pass thresholds or tipping points and become newly fashionable. It seems that low carbon systems will only become significant if low carbon lives become fashionable on a global scale (as argued in Urry, 2011).

Society also matters because people organise, often in and through the media and new media. They seek to preserve, extend or develop relations with others as they fight for or against aspects of their changing environment. Sometime this involves organising for the local as with the Transitions Towns/Cities Movement, sometimes to generate national/European policies and sometimes connecting with others across the globe, as with the thousands of NGOs present at Rio and other Earth summits. Often these actions are intertwined with the perceived arguments of science and people's understandings of the physical world (see Hulme, 2009).

Thus this chapter argues that the 'social' should be positioned at the heart of both why climates are changing and of policies concerned with developing low carbon alternatives. It especially demonstrates how high and potentially low carbon social practices are organised into powerful 'socio-technical' systems. First though, I consider some different academic, social and political responses to the issue of climate change and how these have been analysed by social science.

Responses to 'climate change'

Broadly speaking we can identify three broad positions or discourses within the climate change literature. *Gradualism*, as represented in the reports of the IPCC, involves the claims that climates are changing around the world, that human activities are significantly responsible for these changes, that these changes are relatively slow, and that economies need to be adjusted in order to reduce future temperature increases (see Shackley, 1997, on the IPCC). Individuals and societies can and should be induced to transform their behaviour through appropriate incentives, as elaborated for example in the Stern Review (Stern, 2007). This also presupposes developing new technologies that will somehow fix the problem of climate change by introducing new ways of generating energy.

The organised 'power of science' of the IPCC is distinct in its worldwide mobilisation to deal with the world crisis of global climate change. The IPCC-organised actions of thousands of scientists, policy makers and NGOs across the globe have, in the face of huge commercial and state interests, transformed public and policy debate from around 2005, if not yet becoming global public policy. The IPCC is the world's largest ever scientific endeavour and one remaining relatively open to industry experts, outsiders and NGOs. IPCC has mainly deployed models and arguments that *all* involved can sign up to, with even the Pentagon arguing that climate change will result in a global catastrophe costing millions of lives with its threat to global stability far eclipsing that of global terrorism (see Abbott, 2008). Alongside the IPCC there developed a huge climate politics, in science, in the media including Nobel prize-winning movies (made by former US Vice-President Al Gore) and in much policy debate involving most major global institutions that signing up to the notion of enhanced 'sustainability'.

The second main position is that of *scepticism,* which involves challenging the sciences of climate change especially in light of the huge uncertainties involved in predicting changes in temperatures over future decades. It is said that there are too many 'unknown unknowns'. Also if climates have altered in the past this was argued to be the result of 'natural' processes such as sun spot activity rather than processes which are 'anthropogenic'. Scepticism thus involves a critique of the social sciences playing any role here.

This scepticism is particularly significant within the internet, the blogosphere and many American thinktanks that combat climate change science in order to promote 'business as usual'. These thinktanks include the American Enterprise Institute, Americans for Prosperity, Cato Institute, Competitive Enterprise Institute, Energy for America, Global Climate Coalition, Heartland Institute, Marshall Institute, the Nongovernmental International Panel on Climate Change (NIPCC), Science and Environmental Policy Project, Science and Public Policy Institute, The Heritage Foundation and World Climate Council. Many of these are in effect 'front' organisations intended to suggest that there is more uncertainty about climate science than there actually is. The significance of such climate sceptic 'merchants of doubt' has been well-documented (see Oreskes and Conway, 2010).

There are also some major texts which deconstruct climate science such as the supposed hockey stick pattern of temperature rises over the past millennium (see the critique in Montford, 2010). Some sceptics seek to explain away arguments for climate change as being driven by the vested interests of research scientists and the media, something reinforced since so-called Climategate involving the analysis of emails exchanged between climate scientists, especially those working at the University of East Anglia. The power of scepticism has recently grown, especially in the US. While only one climate change scepticism book appeared in 2001, eighteen were published during 2009.

A different line of sceptical argument is presented by political scientist Lomborg, who argues against the cost involved in dealing with climate change, as compared with other equally important global challenges (2001, 2008). He argues that many actions being considered to stop global warming will produce a lower return compared with other interventions and they may anyway impact little on the world's temperature for some centuries. Other sceptics suggest that there will actually be benefits from changing climates while some migration of populations happens anyway. A connected argument is that of China and other developing countries which claim that the issue of climate change must be mainly dealt with after something like western levels of economic development are secured within their societies (see Watts, 2011, on climate change debates in China).

Social science researchers have explored the changing impact of such scepticism. McCright and Dunlap explore how the American conservative movement sought to protect the notion of unfettered modernisation and to undermine the gains of environmentalism that had been achieved in the early 1970s, with 1970 itself being classified as the 'year of the environment' (2010). Climate change scepticism critiqued progressive movements and especially environmentalism by partly using techniques that had been developed by environmentalists in the 1960s and early 1970s. McCright and Dunlap show how American conservatism cast doubts over the claims of climate change science, manufacturing much 'uncertainty' and undermining potential international cooperation.

Catastrophism critiques both these positions. It takes from the former a belief in the reality of climate change and from the latter the significance of uncertainty and the limits of science. But it then locates both of these within a 'complexity' framework which emphasises non-linearity, thresholds and abrupt and sudden change. It is argued that IPCC Reports do not factor in all the potential and uncertain feedback effects such as the potential contribution of rapid melting processes in Greenland and the Antarctic. These changes in ice are relegated to a footnote in the

Fourth IPCC Report in order to achieve a 90% certainty (Yusoff, 2009). Very modest projections of sea level change, which ignore future uncertainties especially related to the melting of ice, enable sceptics to argue that such increases can be dealt with through modest techniques of adaptation (see critiques of IPCC in Hansen, 2011).

Catastrophism draws upon historical, ice core and archaeological data to maintain that positive feedbacks will take the climate system away from equilibrium through positive feedback effects. Many scientists also generated the thesis that the earth is a single complex system and can be subject to very rapid system shifts moving abruptly across thresholds. Deploying such analysis of climate forcing, Lovelock refers to the likelihood of irreversible global 'heating' (2006).

Thus Wynne suggests that far from the IPCC exaggerating the dangers of anthropogenic climate change they probably underestimate it (2010). Climate scientists tended to present future climate change as gradual and manageable, and underplay the possibility of abrupt changes and runaway feedback loops. Giddens suggests that climate change differs from any other problem faced today, since if unchecked the consequences will be catastrophic for human life on earth (Webster, 2009).

We can finally note here that any description and prediction of climate change and its impacts is entangled with specific imaginaries of how society is and how it ought to be. Even the most apparently technological of proposals will carry with it certain ideas of the social and good society.

The problem of science

The social sciences and especially social studies of science (STS) have examined some of the complex characteristics of science as a system. STS analyses of the physical sciences have shown the way in which the 'unequivocal' truth of the changing of the climate has been established. Such studies also show how climate change came to be the paramount environmental problem rather than many of the other environmental issues discussed since the 1960s and the founding of modern environmentalism that followed Rachel Carson's iconic *Silent Spring* (1962).

These STS analyses also show that the very reliance on complex, highly specialised 'big science' creates its own problems for the politics and policy of climate change. The General Circulation Models used to predict rates of GHGs and temperature increases contain very many 'unknowns', what Keynes called uncertainty as opposed to risk (1921). These arise because of the array of interdependent complex systems involved here and in modelling some fearsomely uncertain futures. Most predictions of future temperature increases through General Circulation Models are unable to factor in multiple physical and social feedback mechanisms, especially those which are yet to occur (Pearce, 2007). And yet climate science is meant to deliver robust predictions for policy makers although the 'science' is still being formed and, in terms of some of the main data, is less than fifty years old. Moreover, most of the science is not laboratory-based – the earth itself is the laboratory.

Also many different 'sciences' are involved in determining the nature of changing climates and they pursue different theories, methods and types of result. Especially significant has been the paradigm shift in climate science at around the turn of the century. Linden notes how as a result of this new research programme: 'the climate community was in the midst of a full-blown paradigm shift' from about the year 2000 (Linden, 2007: 27). This new paradigm concerned exploring new techniques to determine what happened in previous periods of dramatic climate change on planet earth in order to imagine what may well happen in the future.

Alley showed in his research on Greenland ice cores that half the warming between the last ice age and the subsequent postglacial world of around 9 degrees Celcius took just one decade (see Pearce, 2007). This was an astonishing rate of warming. If a similar rate of increase were to happen in the next few decades it would totally transform most forms of life on earth. This ice core research programme shows that there seem to be only two states of the earth's climate, either an ice age or a relatively warm interglacial age. There is no evidence of gradual movement from one state to the other. This ice core research shows how in the past there have been sudden abrupt jumps as the earth responds to carbon shifts with what Pearce calls 'speed and violence' (2007).

STS analyses of the climate sciences show that climate change is not a single 'cause' or a single set of 'effects'; it is sets of changing probabilities. And as Hulme argues, climate change is not the sort of thing that can be 'fixed' in any straightforward sense with a single solution, as compared with the 'hole' in the ozone layer generated by CFCs which could be more or less fixed (2009). Indeed there are very many disparate elements of changing climates: increase in Arctic/Antarctic temperatures faster than that of the earth in general; reduced size of many icebergs and the melting of icecaps and glaciers; reduced permafrost which could release methane in vast quantities; changes in rainfall especially as the sea heats up faster than the earth; reduced bio-diversity; new wind patterns; fiercer droughts and heat waves; more intense tropical cyclones and other extreme weather events; and catastrophic problems for certain iconic species such as polar bears (Yusoff, 2009; generally here, see Hansen, 2011). Interestingly the insurers Swiss Re estimate that losses from these weather events have risen five-fold since the 1980s. Oxfam reports that while earthquake numbers remained relatively stable there has been an almost three-fold increase in flooding and storm events. As environmental journalist Vidal writes: 'Warning: extreme weather ahead', available at ⟨http://www.heatisonline.org/weather.cfm and accessed 24 December 2011⟩.

Moreover, states and corporations are often 'vulnerable' and unable to cope with such fast-moving and unpredicted disasters. These disasters include potential oil shortages, droughts, heat waves, flooding, desertification, mobile diseases and the forced movement of up to 150 million environmental refugees that is predicted by 2050. There are many instances where 'states', relief organisations and companies cannot deal with complex and unpredicted disasters which necessitate improvising new kinds of unexpected mobilities and systems. Two recent examples show this: New Orleans in the US in September 2005 and the nuclear collapse in Fukushima in Japan caused by the massive tsunami on 11 March 2011. These events demonstrate how most organisations cannot deal effectively with failures in institutions and processes as they impact upon each other. Examining such emergencies and crisis responses is becoming a major task for social science research.

Especially significant are negative synergies occurring within the belt around the centre of the earth between the two tropics. The book *Tropic of Chaos* powerfully examines the resulting consequences of anthropogenic climate change and especially extreme weather events already occurring within this huge area (Parenti, 2011). There are many 'damaged societies' experiencing water and food shortages, rising sea levels, poverty, lack of access to energy, refugees, extreme weather and regime failure. They compound and amplify each other through a kind of 'catastrophic convergence'. Up to 2.7 billion people are likely to experience violent conflicts as climate change interacts with these other system contradictions (Parenti, 2011: 7–11).

Moreover, women within such societies will be especially affected by such catastrophic convergence, available at ⟨http://www.un.org/ womenwatch/feature/climate_change/ and accessed 8 January 2012⟩. If there is mitigation funding it mostly supports large-scale clean technology projects rather than the smaller scale projects benefiting poorer people and especially women farmers. There is a danger of developing under the cloak of 'sustainability', masculinist 'green monumentalism' such as high-speed train lines or huge geo-engineering projects which in their construction and operation generate very high levels of GHG emissions. It is local knowledge and technologies which should be of greater importance, and they are more likely to be developed by and relevant to women's interests.

Politics of climate

Protesting against high carbon forms of life and its resultant carbon emissions is historically unusual and difficult to bring about. Most politics historically has been *against* the income, wealth and power of other social groups, against their 'goods'. Such conventional politics involves the desire to effect some reform or a revolution in the local or national distribution of such goods. Protesting social groups typically seek more goods for themselves, either in the short run through reform, or in the longer run through a revolution in the overall system of distribution. But something different started to happen to 'western politics' in the 1960s. What could be described as an archipelago of movements began to emerge more or less simultaneously. These groups, sometimes described as post-industrial or post-affluent, sought to bring about a different transformation and politics. Here as noted above, the emphasis is upon bads and upon how aspects of these bads affect everyone, albeit variably. These bads are manmade, incalculable, uninsurable against and sometimes catastrophic in their potential scale (Beck, 1992).

These bads are not accidental or incidental to forms of life. They stem from society and thus society itself needs to be reformed or revolutionised in order to minimise such bads, both now and especially in the future. It came to be widely argued during the 1970s that future generations were part of a given society and should not be viewed as less important or less worthy members. The interests of future generations, to use the language of economics, should not be subject to any discounting. One component of this 1960s radicalism was that society should be practised as multi-generational. The interests of current generations should not be exercised at the expense of yet-to-be-born generations. This notion of intergenerational sustainability was enshrined in the 1987 Brundtland Report that is presciently entitled *Our Common Future*, that stretches into and across present and future generations.

And when the struggles against climate change began to develop it became even clearer just what a strange politics is involved here. Climate change politics involves campaigning not for abundance or growing abundance but for reduced abundance now so as to ensure reasonable abundance in the long term for future generations and in other parts of the globe. Unlike almost all other politics in world history it is a politics for lower consumption of goods and services now. This is a politics directed against others but especially against oneself and the high carbon systems that historically made life in the rich North nice, comfortable and long.

Because this campaigning is for reduced consumption bringing that politics into being on a mass scale across various societies is enormously difficult to achieve. It is in some ways astonishing that since the 1960s some progress has been made in developing discourses and practices that do articulate in quite complex ways some forms of intergenerational and international interdependence. How has this notion of interdependence partially come about, especially given the power, reach and effectiveness of the carbon interests, as well as the continued organisation of national states pursuing narrow short-term interests?

Various commentators deploy a notion of the cosmopolitan to characterise the increasing mundane interconnections between peoples, places and societies. As Beck brings out, this notion of cosmopolitanism does not entail global uniformity or homogenisation (2006). Rather, groups, communities, networks, political organisations, cultures, and civilisations remain diverse but with some of the walls between them getting replaced by bridges. These bridges are erected in various imaginaries ('cosmopolitan visions'), in nations and localities, in systems of norms, in institutions, as well as within global politics.

What then are the connections between such cosmopolitan bridge building and changing climates? Hulme draws out three aspects (2009). First, an increasingly cosmopolitan science and media (and we add travel) have helped generate the understanding of climate change as being *global* and as *the* environmental issue of the day. NGOs likewise develop and deepen such understandings of climate change through their cosmopolitan social practices. And indeed people increasingly experience the weather elsewhere through travel, webcams, the internet, TV and so on. Weather elsewhere is routinely available and known about. It is both strange and cosmopolitanised. Second, the very notion of climate change dissolves many boundaries and helps to create something of a common world. There is less sense of an 'outside', an 'exit' or the 'other'. It thus helps to extend and deepen cosmopolitanism as a social force. Climate change contributes to cosmopolitanism through the ways in which its science, politics and media generate new kinds of mobility, openness, reflexivity, plurality and public spheres. Third,

Hulme says there is no such thing anymore as a stable 'natural' climate to which the current world climate will somehow return if only GHG emissions would cease. And part of such cosmopolitanism is to recognise that all climates are hybrid mixtures of natural and social worlds. The social is centrally significant in developing something of a future cosmopolitanism.

Low carbon systems

Finally, I consider how contemporary societies got into this 'catastrophe' in the first place. In the fateful twentieth century a cluster of high carbon systems became sedimented. Societal changes brought about high carbon forms of life, as well as high population growth. Especially important in this patterning from the twentieth century were a cluster of carbon-based systems beginning in the US in the first half of the last century, including electric power and national grids; the steel and petroleum car system; suburban housing; technologies for networking; distant, specialised leisure sites; and aeromobility. These high carbon socio-technical systems increased income, wealth and movement, engendered population growth, generated GHG emissions and used up maybe half of the world's oil that made this world go round (see Geels et al., 2012 on the automobility socio-technical system).

And in the neo-liberal period since the later 1970s there was a powerful ratcheting up of such systems with many generating a striking 'excess' of energy, choice and addiction. Various places of excess developed in multiple locations around the world, with Dubai setting the standard for exceptional producing and consuming to excess (Urry, 2013). The legacy of such excess during the twentieth century can be seen in the limited future alternatives that are now possible for twenty-first century societies which have been dealt a weak hand for developing post-carbon societies.

In order to overcome the problems of this high carbon world it is necessary to bring about a wholesale shift to an interlocking *cluster* of low carbon systems. This involves establishing a low carbon 'economy-and-society'. This is not just an economic change here. The utterly crucial need is for moving to interlocking low carbon systems but which may provide lower levels of measured income and population levels but sustain reasonable levels of wellbeing or what Jackson terms 'flourishing' (2009). This is not at all simply a matter of policy prescription or of transformed economic incentives, but of different patterns of social life in most domains within most societies.

Within a single society, such a low carbon shift is more likely it seems, the more equal the society in question, the greater the scale of local social experimentation, the more that decisions can be taken locally or at least nationally, and the greater the finance, human capital and social capital that can be put into post-carbon programmes, initiatives and experimentation (Wilkinson and Pickett, 2009). Especially important is the developing of vanguardist thought and policy in 'low carbon' which is not privately owned but becomes part of 'a global low carbon commons' (see Geels et al., 2012, on the possibilities of this with regard to low carbon transport).

If there is no move to a low carbon economy and society within most high carbon countries and cities within a couple of decades, then many places through changing climates and the relative or absolute decline in oil and gas supply (with resulting consequences for food and water) will develop major negative system effects. These may include greater levels of ungovernability, major losses of income and population, and large increases in personal and system surveillance.

Changing climates, rising world population (from seven to more than nine billion), and declining oil will transform the resources underlying societal formation and reproduction even in the rich North. There will be variations across societies in the significance attached to climate change, in the degree of acceptability of different kinds of change, and in the willingness to seek to regulate and govern such uncertain futures. Climate and societal futures seem similarly beset by huge 'uncertainties' as opposed to calculable risks.

Change from the 'top' might occur but through a 'shock doctrine' and a 'global war' that short circuits normal procedures and protocols (as happened after September 11 with the 'global war on terror'). A massive collapse of oil supply or oil price increase or prolonged flooding or drought in a global city could provoke a 'war on climate change'. This presupposes that global science, politics and media all successfully frame this 'event' as bound up with and contributing to 'climate'. Shock doctrines and many kinds of 'war' are generally bad for any enduring democratic practices. Developing a low carbon 'economy and society' might result through such a shock and a global war on climate. But this would be a corporatist, top-down surveillance low carbonism as opposed to a more localist, decentralised self-organising low carbonism. And both seem exceptionally hard to realise.

Thus this is not a question of changing what individuals do or do not do but changing whole *systems* of economic, technological and social practice. Systems are crucial here and not individual behaviour. And systems are not just economic or technological but they also presuppose patterns of social life which come to be embedded and relatively unchanging for long periods, such as the 'steel and petroleum' automobility system during the twentieth century. Such high carbon systems have got into social life. And that is why I refer to the need to reverse 'economy and society'.

We have seen how systems form habits and these habits are the stuff of social life and not easily changeable, and certainly not by states, in which people have low trust, instructing them to change. Changing systems must appear to populations around the world as more desirable, fashionable and necessary components of a better and more fashionable life. And these topics are utterly significant issues for very many branches of the social sciences.

Without some extremely rapid transformation to a cluster of low carbon systems, then we can anticipate, along with urbanist Mike Davis, that by 2030: 'the convergent effects of climate change, peak oil, peak water, and an additional 1.5 billion people on the planet will produce negative synergies probably beyond our imagination' (2010: 17). With all this in in mind, it is dear that good social science and effective low carbon design are both urgently required.

Bibliography

Abbott, C. (2008). *An Uncertain Future*, Oxford: Oxford Research Group.
Beck, U. (1992). *Risk Society*, London: SAGE.
Beck, U. (2006). *Cosmopolitan Vision*, Cambridge: Polity.
Carson, R. (1962). *Silent Spring*, New York: Fawcett Crest.
Davis, M. (2010). Who will build the ark? *New Left Review*, 61, 10–25.
Geels, F., Kemp, R., Dudley, G., and Lyons, G. (eds) (2012). *Automobility in Transition?* London: Routledge.
Giddens, A. (2009). *The Politics of Climate Change*, Cambridge: Polity.
Hansen, J. (2011). *Storms of My Grandchildren*, London: Bloomsbury.
Hulme, M. (2009). *Why We Disagree about Climate Change: Understanding Controversy, Inaction and Opportunity*, Cambridge: Cambridge University Press.
IPCC (2007). *Climate Change 2007: Synthesis Report. Contribution of Working Groups I, II and III to the Fourth Assessment Report of the Intergovernmental Panel on Climate Change*, Geneva: IPCC.
Jackson, T. (2009). *Prosperity Without Growth*, London: Earthscan.
Keynes, J. M. (1921). *A Treatise on Probability*, London: Macmillan & Co.
Kirk, G. (1982). *Schumacher on Energy*, London: Jonathan Cape.
Linden, E. (2007). *Winds of Change. Climate, Weather and the Destruction of Civilizations*, New York: Simon and Schuster.
Lomborg, B. (2001). *The Skeptical Environmentalist*, Cambridge: Cambridge University Press.
Lomborg, B. (2008). A new dawn, *Wall Street Journal*, 8 November.

Lovelock, J. (2006). *The Revenge of Gaia: Why the Earth Is Fighting Back – and How We Can Still Save Humanity*, London: Allen Lane.

McCright, A. and Dunlap, R. (2010). Anti-reflexivity: the American conservative movement's success in undermining climate change science and policy. *Theory, Culture and Society*, 27, 100–133.

Montford, A. (2010). *The Hockey Stick Illusion. Climategate and the Corruption of Science*, London: Stacey International.

Oreskes, N. and Conway, E. (2010). *Merchants of Doubt*, New York: Bloomsbury.

Parenti, C. (2011). *Tropic of Chaos*, New York: Nation Books.

Pearce, F. (2007). *With Speed and Violence. Why Scientists Fear Tipping Points in Climate Change*, Boston: Beacon Press.

Shackley, S. (1997). The intergovernmental panel on climate change: consensual knowledge and global politics. *Global Environmental Change*, 7, 77–79.

Shove, E. (2010). Beyond the ABC: climate change policy and theories of social change. *Environment and Planning A*, 42, 1273–1285.

Shove, E., Panzar, M. and Watson, M. (2012). *The Dynamics of Social Practices*, London, SAGE.

Stern, N. (2007). *The Economics of Climate Change: The Stern Review*, Cambridge: Cambridge University Press.

Urry, J. (2011). *Climate Change and Society*, Cambridge: Polity.

Urry, J. (2013). *Societies Beyond Oil*, London: Zed.

Watts, J. (2011). *When a Billion Chinese Jump*, London: Faber and Faber.

Webster, M. (2009). Uncertainty and the IPCC. an editorial comment. *Climatic Change*, 92, 37–40.

Wilkinson, R. and Pickett, K. (2009). *The Spirit Level: Why More Equal Societies Almost Always Do Better*, London: Allen Lane.

Wynne, B. (2010). Strange weather, again: climate science as political art. *Theory, Culture and Society*, 27, 289–305.

Yusoff, K. (2009). Excess, catastrophe and climate change. *Environment and Planning D. Society and Space*, 27, 1010–1029.`

4

Waste, Resource Recovery and Labour: Recycling Economies in the EU

Nicky Gregson and Mike Crang

Introduction

Waste and its connections to a profligate consumerism have become a core concern for public policy in the developed nations in the twenty-first century. Whether the focus is food or mobile phones, computers or electrical goods, waste – and particularly post-consumer waste – is never far from view and moral concern. Reports have shown how at least a third of the food produced by the planet ends up not on the plate but as food waste (WRAP, 2008; Stuart, 2009; IME, 2013). Trends to faster manufacturer product cycles, cheaper products undermining the economics of repair, and technology and fashion-induced pressures to upgrade goods are seen to be critical to the emergence of 'a throwaway society' (Packard, 1960; Strasser, 2000; Cooper T. H, 2005, 2010; see also Gregson et al., 2007a). In the EU it is accepted that this has to change. A wave of policy initiatives has identified minimising and preventing waste as central to sustainable futures.

In waste policy, a first priority is minimising waste generation; the second is to reduce waste by diverting unwanted materials from the waste stream. Diversion is achieved via recovery, be that for reuse, recycling or as energy from waste. Recovering wastes via recycling means that wastes become potential feedstock for other industries: they have become resources. This is a major transformation – not just in how we think about waste, but also in what happens to it. Disposal through 'controlled tipping' (or landfill) has gone from being the default option for waste in many EU-member states to becoming the least desired option for managing wastes. This is not only because of the contribution of landfill gases (principally methane) to greenhouse gas emissions and hence to climate change, but also because materials disposed of as wastes are no longer circulating and so are lost to economies whilst their replacement depletes natural resources.

Since the late 1990s the imperatives to minimise and reduce waste have been enacted through the principles of the EU-wide Waste Hierarchy.[1] As a result, national level waste policies have implored millions of European households and consumers to Reduce, Reuse and Recycle, promoting these actions as part of a transition to sustainable consumption. In the UK, long stereotyped as the 'dirty man of Europe', this has necessitated the development of an infrastructure to divert waste materials from landfill. This is the biggest change to have occurred in domestic waste management in the UK since the end of the Second World War (see also Cooper, 2010). Rather than one bin for 'the rubbish', UK households now have multiple bins for multiple materials: dry recyclables (paper, card, tins and plastic containers), glass, green waste, food waste and residual waste. Increasingly too, the principles of the Waste Hierarchy are being extended to producer responsibility. Manufacturer and retailer responsibilities with respect to waste minimisation and recovery are being ratcheted up, whilst 'end-of-life' is a core principle in sustainable product design.

In policy terms, the Waste Hierarchy aligns with the Waste Framework Directive. Things that are discarded become classed as wastes. Once classified as a waste, considerable restrictions are placed on the movement of things.[2] Wastes require permits to enter markets and to be moved and exchanged. Producer responsibility here is correlated with a 'proximity principle', where, rather than exporting harm, wastes should be handled by those who made them; ideally near to where they are generated – not distanced physically and psychologically by being sent out of sight and out of mind to be 'somebody else's problem' (Clapp, 2001).

Environmental values underpin EU waste policy. Restrictions and regulations seek to prohibit global flows of wastes, primarily to the Global South and to China. They seek to prevent the dumping of hazardous materials on people and environments lacking the facilities, technologies and regulatory context to manage them appropriately. Inspired by environmental justice movements, these values attempt to corral wastes closer to home.

In the social sciences, research on waste has focused overwhelmingly on the governance of (municipal) waste and preventing harm (e.g., Davoudi, 2000, 2009; McDonald and Oates, 2003; Tonglet et al., 2004; Davoudi and Evans, 2005; Bulkeley et al., 2007; Davies, 2008).

Other research has highlighted how consumer-derived waste connects with social relations (Chappells and Shove, 1999; Gregson et al., 2007b; Alexander et al., 2009). How might new infrastructures and habits of recycling and/or disposal be adopted, and by whom? Why are patterns of sorting and disposal that are normal in Scandinavia so difficult in the UK for instance? However, somewhat surprisingly, there has been little work on waste's transformation to resources. There has been precious little interest on the part of social scientists in following recovered materials back into economies (Alexander and Reno, 2012).

In academia, the challenge of turning wastes to resources is one which has been taken up largely by the engineering and technical disciplines, where the focus is on technical possibilities – or what physical transformations can be wrought through chemical and/or mechanical processes.[3] This will not suffice. To focus exclusively on the technical is dangerous for, as the social sciences caution, what is technically doable is not always socially and economically possible. Take the example of Shijiao, a town in China that is the world capital for Christmas tree lights recycling (Minter, 2011). Each year roughly 20 million pounds (lbs) of discarded Christmas tree lighting is processed by the factories in this town, much of it from the US. Whilst it is technically possible to shred wire anywhere in the world in order to recover it for recycling, the trouble is that there is little demand from manufacturers in the US for this recovered material. In contrast, there is demand in China, where it is not just the recovery of copper wiring that is economically viable but the recovery of insulation too. Here, plastic insulation from Christmas tree lights has become the raw material for manufacturing slipper soles. Recovering wastes, then, is not simply an environmental and technical issue. Critically, it depends on the economic geographies of manufacturing, or – more simply: 'It all depends on what people are making, and right now, in the US, they really aren't making much with that kind of mix' (Minter, 2011). Made in China, therefore, tends to mean Recovered in China.

In The Waste of the World programme of research[4] we have followed wastes back into economies. We have charted and analysed some of the contours of global recycling economies and shown how wastes generated in the EU are recovered and revalorised beyond the EU, often in conditions of minimal environmental and labour regulation. Some of these trades are, at best, 'grey', at worst plain illegal. But they go on nonetheless, because they are immensely profitable (for some), there are endless regulatory loopholes or 'fudges,' and the impossibility of checking every container box destined for export, means that what is classified as a waste can easily morph to be recategorised as reusable goods (Kamuk and Hansen, 2007). This, notwithstanding the WEE directive forbidding the export of scrap electrical goods, means that used but allegedly still usable televisions are one of the biggest exports from the EU to Ghana, for example (EEA, 2009), where the vast majority are then broken up for recycling. Tracking the trades in used textiles and end-of-life ships has taken us to parts of South Asia, specifically to Panipat in northern India and Chittagong in southern Bangladesh respectively.[5] At the same time, our work has examined aspects of the increase in recovery-for-recycling work within the EU promoted by the collection of materials for recycling and the push to recover materials closer to home. New materials recovery processing factories are appearing apace across the EU, with more proposed. The rhetoric which accompanies them is that they are creating new forms of 'green' employment.

In this chapter we draw on three of our research projects, focused on end-of-life ships, textiles recovery and 'dry' household recyclables, to highlight the realities of these new 'green' jobs in the EU's new 'green' economy.[6] Drawing on an old critical social science mantra, of rendering apparent that which is otherwise hidden, our work shows that, even within the EU, 'green' is a very long way from 'clean', and that the rise of recovery-for-recycling activities can be firmly located within issues of low paid as well as transient and often migrant work. We then turn to consider the products of these EU recovery activities. In this chapter we focus on UK-based recovery, highlighting connections to global export markets, and in conclusion we raise some fundamental questions about the EU's future as a recycling economy and society, and reiterate the importance of further social science work which recognises the challenges economies pose to the policy drive to create sustainable futures.

Green, but not clean: recycling work

By 3pm Abdul was facing down the bottle caps, a major sorting nuisance. Some had plastic interior linings, which had to be stripped out before the caps could be assigned to the aluminium pile. Rich people's garbage was every year more complex, rife with hybrid materials, impurities, imposters. Planks that looked like wood were shot through with plastic. How was he to classify a loofah? The owners of the recycling plants demanded that waste was all one thing, pure. (Katherine Boo, 2012)

The Abdul figure of the Annawadi slum next to Mumbai's international airport is a familiar one in the social science literature, in which rag-pickers and scavengers located in the Global South eke out a living from the garbage discarded by affluent city dwellers (e.g., Asim, Batool and Chaudhry, 2012). Their marginal existences are indictments of global capitalism and its inequalities. Existing on incomes that are below the global poverty line, it is their livelihoods amidst the waste and detritus generated by affluence and consumerism, and in conditions of minimal environmental regulation, that command attention for western eyes. They are iconic workers, for the very reason that they depict the dark side, or black sun, of globalisation. Yet they are also so often depicted with a large dollop of exoticism as seemingly miraculously recovering value from rubbish and enacting a recycling economy. By drawing attention to the practices of Abdul's work, however, Boo goes beyond this figuration to highlight the tasks of classification, segregation and sorting which characterise the work of materials recovery. Materials recovery precedes recycling. For recycling processes, purity matters. Correspondingly, recovering materials for recycling is an activity in which value is realised through the labour of classification, segregation and sorting. More segregation and sorting yields higher grades of material, more purity and less contamination. Conversely, minimal segregation and sorting results in higher levels of contamination and less desirable recyclates. These principles apply wherever materials recovery is done.

In much of Africa, Asia and South-East Asia, segregation and sorting work is still largely manual work, if no longer performed predominantly by scavengers. Factories segregating and sorting used goods and recovered materials have become increasingly commonplace. Whilst there is some mechanisation of the labour process, cheap labour results in continuing high levels of manual work. In contrast, in western European countries in the EU, recovering materials for recycling is more highly mechanised. Capital intensive equipment designed to process and separate high volumes of materials is commonplace, particularly in the recovery of 'dry recyclables' (paper, aluminium, glass and plastics). Nonetheless, manual work remains a critical part of operations and, in some sectors, notably textiles, continues to be the primary means to sorting and segregation. What is this recovery work like?

Inside the UK's materials recovery facilities (MRFs), the contents of dry recyclables collection vehicles are loaded mechanically into bays and then placed on a conveyor belt, to enter the first stage of processing: a 'picking cabin'. Here teams of six to eight, mostly men, stand on both sides of the belt wearing hi-visibility vests; most also wear ear defenders or head phones connected to iPods or MP3 players. The majority wear masks, to counter the smell, and thick gloves to protect against sharp objects (glass, staples, paper). The task at hand is to pull off the belt and place in wheelie bins all those materials that will cause problems for the processing plant and/or which are non-recyclable. For the most part this is stuff like over-sized card and the wrong sorts of plastics but it can also be stuff that should not be in the recyclables stream – like hospital waste, dead animals, wheelie bins, paddling pools, car wheels, or anything else that gets placed in recycling bins by publics that shouldn't be. Shifts are long – typically 8–11 hours. Most MRFs work 24/6, with one day for maintenance, whilst a few work 24/7/365. Job advertisements for 'pickers' specify the need for 'attention to detail' and the ability to stand for long periods of time. They neglect to mention the noise, intense heat and smell that are a feature of the work. The pay is minimum wage.

In textile recovery factories yet more teams, called sorters or graders, stand by multiple conveyor belts or by sorting tables. The loads that arrive here are sourced from charity shop rejects or unsold goods, textile recycling banks and door-step collections. Most employees here are women. Typically one large factory would employ around 100 sorters at any time. Unlike their counterparts in the MRFs, their task is to sort and grade used garments, initially by quality ('best', 'useful', 'worst') then by recyclable/reusable; by material characteristics; by garment type and season. The sub-categories are numerous, as well as closely guarded for commercial reasons. Publicly available information lists some 20–30 categories, with seven separate categories of women's blouses, for example: cotton, denim leisure, designer sleeveless, polyester, silk, white cotton and white silk. The reality is more of the order of 70+ categories, with some factories having as many as 160.

The labour process of textile recovery work echoes that in an MRF picking cabin. It involves continual standing at conveyor belts with minimal breaks; the smell of used clothing is pervasive and contaminants include blood, vomit and excrement. Many sorters wear dust masks, provided by the factory, but – unlike in MRFs – textile recovery factories forbid the wearing of gloves. This is because sorting and grading are dependent upon the accurate assessment of material qualities. These qualities cannot be evaluated by look alone, but rather require touch and feel to discern. Allergies – nasal and skin – are common amongst workers. Shifts are again long – typically 8–5 for full-time employees. The work is extremely low paid. For most sorters it is no more than minimum wage, and there is no career or pay progression. Job descriptions and advertisements appear rarely, for most work is acquired via social networks. In one factory we studied, Russian was the lingua franca, and the women working on the lines mostly came from Eastern Europe, principally Lithuania, Bulgaria and Russia. In previous years, the work force had been dominated by West Africans, by workers from the Caribbean, and before then by workers from Pakistan. In each labour market phase, prevailing ethnicities relate to their perceived knowledge of key international markets in second-hand textile, and much of this labour is currently Eastern European. Once employed, 'lock-in' to the work is common. Long hours and constant standing result in tiredness, an inability to search for other jobs, and low self-esteem. In the UK, the sustainable economy of recycling has led to the return of the type of factory assembly line work that disappeared with the shift of assembly based manufacturing to the Far East in the 1970s and 1980s. Like their counterparts in assembly, which a generation of classic sociological studies made visible, these new green jobs are physically tough and monotonous work.

It is not just consumer goods but also capital goods that need recycling and offer valuable recovered materials. End-of-life ships, trains and aircraft are valuable in terms of the metals they yield for recycling, notably aluminium, copper and ferrous scrap. But to pull these complex things apart is complicated. Usually their disassembly involves dealing with hazardous materials, as well as salvaging different kinds of materials such as textiles (seat covers, carpet) and plastics. It also involves separating ferrous, from non-ferrous metal; then segregating the non-ferrous into key metals and alloys (e.g., copper, aluminium, nickel, manganese). The physical science principle sounds simple but the practice is anything but. For, just as Abdul recognises (and bemoans), the problem lies with the heterogeneity of these manufactured things. So, in the case of end-of-life ships and trains, the legacies of past manufacturing practices mean that valuable copper (the conductor of electricity) is almost always surrounded by sheeting (which, if mistreated, releases toxic chemicals). To get at one requires the safe removal (and disposal, at cost, in a landfill) of the other. Unlike textiles and dry recyclables recovery, this type of recovery work is outdoor work, but its milieu is the scrap yard. Whilst the job of some is to cut metal using either the oxyacetylene torch or mechanised shears, others work in dark, confined, negative pressured spaces, lit only by head torches and temporary lighting, stripping out asbestos, or recovering soft furnishings for days on end.

Ship or train recycling may be 'green', through their connections to reuse and recycling, but these activities are certainly not clean. Although compliant with environmental regulation, this type of recovery work is amongst the dirtiest, most physically challenging and hazardous of current occupations in the EU. A migrant division of labour shapes who does these dirty, dangerous and demeaning jobs. A ship and train recycling facility in Northern Europe is indicative. Unlike textile recovery, this is all men's work. The languages spoken span the EU and map into a strongly segregated division of labour that highlights key distinctions within the EU labour market. At the apex of the labour hierarchy are those who cut metal: they are self-employed, French-speaking Belgians. Those who sort metal, however, are Eastern Europeans – Poles and Czechs, hired via an agency on short-term contracts at rates of pay well below those of the cutters. Asbestos removal is performed by another subcontracted company, this time Dutch. Unlike the Poles and Czechs, who return home on a monthly basis, they are weekly commuters – staying in a cheap hotel during the week before returning to the Netherlands at the weekend. They too are employed by-the-job. Job insecurity cuts across all these workers. Transience and itinerancy are characteristics of the lower echelons of this labour hierarchy. Waste may be corralled but it is mobile labour that enables this economy to function. Indeed, the more the work is connected with hazardous wastes and their safe recovery for disposal, the more transient and itinerant the labour is likely to be. An effect is that individual men come and go, unable to stick at the work or its conditions, the continual moving from job to job, and being away from home. A further effect is on occupational health and safety. In these non-unionised and itinerant conditions, issues of long-term occupational health and safety are transferred from companies onto individuals, who have to manage their personal exposure risk. Just as in the heavy manufacturing and mining industries of Europe in the early-mid-twentieth century, short-term financial gain counter-balanced against longer-term occupational health is an issue for all who work in the new metal recovery sector. The picking cabins of MRFs across the EU raise similar issues of occupational health. Remarking on a tour of a model German facility, a senior manager in a UK local authority municipal waste management team recounted the horror of seeing the reality of a picking cabin in a supposedly state-of-the-art facility: 'guys – all of them Turkish – stripped to the waist, lathered in sweat, and working in 90 degrees Celcius for 11 hours a day'.

Across the EU, the emergence of new green recycling jobs translates to a reality of highly segregated, low or minimum wage labour markets, characterised by physically challenging and often hazardous work, and where migrant, transient and itinerant workers are often to be found. This labour is the means to value creation. Value itself, however, is produced through physically segregating and sorting materials or things into categories. Segregation and sorting is the point at which value is recovered (or extracted) from waste, which is the unsorted residue that remains. The labour process in recovery-for-recycling rests on multiple acts of segregation, sorting and classification. Workers' value judgements matter, therefore, to value creation. And where their judgement fails, be it through lack of training or pressure of work, then the recycling economy fails. Examples from our research include workers failing to remove contaminants from MRF conveyor belts, failing to recognise a particular metal, and failing to understand the relationship between the style, look and feel of a garment and its material composition. The effect of such judgements is felt financially and economically. To illustrate the complexly layered judgements and valuations at work in a textile recycling factory, we take the example of the categories 'silk' and 'silky'. If the garment is reusable and saleable in the higher value second hand reuse market then the category 'silk' is inclusive of 'silky' synthetics as it addresses the style and feel of the garment to a possible wearer. However, if it is destined for the lower value materials reprocessing market then synthetic but silky fabric would contaminate that stream where the material purity is paramount. Similarly, failing to distinguish non-ferrous from ferrous metals results in a loss of value, as the higher value non-ferrous is subsumed within the lower value ferrous. Aligning or misaligning categories of quality and materiality can thus drastically alter the value recovered. It is here that the significance of who actually performs the work of materials recovery can start to bite. Employers recognise this. For textiles especially, it is not just a matter of learning to recognise and characterise materials but matching that to the demand characteristic of different export markets. In these circumstances, the competitive advantage of employing migrant workers is not just the classic one of minimising labour costs; it is a way of harnessing knowledge of export markets to maximise the value extracted from the waste stream.

Products and markets

When we get people in and they all ask 'Do you get local companies in to do the recycling locally?' I'm pretty upfront about it. I say, 'Yes we do, if they're economically viable. But if they're not then we've got to go elsewhere.' I just think people ought to be more upfront about that because we've got to survive. (UK MRF plant manager)

Information leaflets produced to encourage UK householders to participate in recycling schemes typically make reference to the local and community benefits of recycling as well as to environmental value. Recycling is represented as acting locally, as creating jobs, as helping local authorities meet statutory recycling targets, and as a way in which we can all save the planet. The reality is more complex. Rather than 'Think global, act local', the pattern might be 'waste locally, recycle globally.' Recovered paper, which is the biggest component by weight of the dry recyclables stream, is among the UK's biggest exports. Most of this material is packed into container boxes and shipped to China. Indeed in 2012 more than 11% of all the UKs exports to China by value were various forms of scrap and recovered materials. It is the same story with many other recovered materials. With used textiles, only 20% of goods collected are resold in the UK, primarily through charity shop outlets (Morley et al., 2009). The rest goes for export into reuse and recycling markets, primarily to Eastern Europe, Africa and South Asia. The rise in recycling, then, has intensified networks of global exchange. There is now more of a closed loop but on a global scale. Consumer goods and agricultural products previously flowed from China, Indonesia, Bangladesh and other countries in the Global South to Europe (and North America), via container vessels that had to then return empty. The same containers are now filled with recovered materials transported at knock down rates on this 'back run', to places which demand those materials.

That the primary markets for many recovered materials are in South and East Asia can be explained by global shifts in manufacturing and assembly. Northern Europe may continue to manufacture high value products in sectors such as pharmaceuticals and defence, but mass market manufacture, particularly in relation to consumer goods, long ago shifted east. MRF operators recognise this. They say: 'Mr and Mrs General Public have a worry about sending stuff to China, but this is where stuff is made and packed now!' Some recovered materials are recycled into products which repeat the journey back to Europe. This is particularly the case with low-grade paper and card, which becomes yet more packaging to wrap yet more consumer products for shipping for onward retailing, before being placed once more in the recycling bin by consumers to start the journey back halfway across the world again. Other recovered materials enter production facilities whose products are destined for the booming domestic markets of Asia. A case in point here is the furniture and soft furnishings that are stripped out of the ships that are broken up on the beaches near Chittagong in Bangladesh.

These materials find their way into numerous furniture remanufacturing outlets, where they are refurbished as domestic furniture products, and then sold by retailers into the homes of middle-class Bangladeshis. The ferrous scrap recovered, which, according to various estimates, amounts to somewhere in the region of 60–80% of Bangladeshi steel production, is rerolled into steel rods and used in the construction industry.

Recovered textiles comprise both reusable goods and recyclable textiles. Reusable goods are sold by commodity brokers into differentiated export markets: Africa, Eastern Europe and Asia. Bales can be 'creme' or standard, and are of different sizes. The Eastern European market, which demands the highest quality, is characterised by smaller bales of goods sold primarily for reuse. In the case of the African market, smaller bales of high quality goods are sold alongside larger bales of goods requiring further sorting. Wholesalers in these export markets typically buy by the container load. They sell on to market traders, who buy by the bale to sell used clothing at local street markets. The Asian market, by contrast, comprises larger unsorted bales with a focus on materials recovery. Rules to protect indigenous clothing manufacture demand that used clothes are lacerated so that they cannot be worn but their fibres can be recycled and then respun into shoddy cloth ending up as blankets, many of which travel on as emergency aid supplies.

That so much recovered material ends up being sold into export markets is not simply a matter of global demand but also how that intersects with labour processes. In some sectors, particularly those characterised by mechanised sorting, this is also to do with the quality of products produced by European sorting and how this relates to the economics of recovering materials, particularly in Northern Europe. With its reliance on hand sorting and a numerically large labour force, textiles recovery in the EU has strong parallels with its counterpart in the Global South. But in the dry recyclables sector in the UK, as we highlighted in the previous section, recovery is as 'lean' as it can be. Labour, at least compared to elsewhere in the world, is at minimal numerical levels, and rates of pay are as low as they can legally be. Notwithstanding this, and plants running at near to maximum capacity, the margins are low. Typically they would be of the order of 10–15%. Further, high capital cost processing equipment requires high volumes of material. In turn, this creates pressures which favour long-term contracts with suppliers, chiefly local authorities, who are equally keen to guarantee that their diversion from landfill targets can be met by MRF operators. What results from this production process is a set of standard, but relatively low-grade products and minimal product innovation. All this has implications when it comes to selling the recovered materials as recyclates.

Recovered materials are only recycled if they can be incorporated into further rounds of manufacturing production. To achieve this requires that recovered materials pass the acceptance criteria of manufacturers. In order to do this they have to meet materials threshold standards, set by manufacturers, which typically are defined in terms of maximum permitted levels of contamination. Contaminants are unwanted materials in the production process. They are things which are either problematic for product quality control, or materials troublesome to processing techniques and technologies. Examples of this from the materials recovery sector would include too high water content in recovered paper, the presence of too much glass and plastic in recovered paper, and the presence of aluminium foil mixed in with aluminium tins. It is here that materials recovery – at least in the UK – runs into difficulties. The UK reliance on co-mingled collections from households means that plastic and paper are collected together. To separate out these two materials streams mechanically to the levels required by both domestic and export markets is challenging. Glass presence is even more troublesome, leading to rejections from the main UK paper mills. Similar problems occur with aluminium containers. Plant processing technology separates all aluminium from ferrous metals. However, the aluminium stream can easily include foil, for example from pet food containers. When unsteam-cleaned, that is designated a contaminant by the food and drink industry, meaning that such recovered aluminium is not accepted by manufacturers producing food and drink containers. Whilst materials recovery from co-mingled collections and MRFs might demonstrate diversion from landfill, the products it results in are often low grade. They struggle to meet the acceptance criteria demanded by many manufacturers, for whom purity matters.

What can be done about this? One answer is better sorting and segregation. That can be achieved via kerbside sorting operations, but doing that requires more labour and more labour time spent sorting. That costs. It also runs counter to the considerable capital investment already sunk into co-mingled collection systems and processing capacity. More finer grained sorting, and reduced contamination, particularly with respect to plastic and paper, can be achieved by running materials twice through a processing facility, but the economics on already tight margins does not add up. Low-grade products look as if they are here to stay. The UK used to be 'the workshop of the world'. Now its MRFs and the recycling infrastructure which supplies them might succeed in demonstrating diversion from landfill but they do so by turning domestically generated dry recyclables into low-grade recyclate for the world.

Conclusions

This chapter has demonstrated that the social sciences are critical to identifying the challenges that face policies that aim to reduce waste through appealing to sustainable futures. Turning wastes to resources cannot be achieved through environmentalism alone; neither is this purely a technical matter which can be left to the engineering disciplines and the 'green' end of the physical sciences. Rather, if wastes are to become resources they have to become products, bought and sold in markets. This means that economies, and particularly the economic geographies of manufacturing, matter – and they matter profoundly to what can be done with what wastes where, why and how.

The EU's policy vision is of the EU as a recycling economy and society. It emphasises endless materials circulation, in which environmentally aware citizens and businesses combine to do the right thing, by managing production and consumption in such a way that there is zero waste, and resources, once extracted, remain circulating within European economies rather than being lost to future economic value generation. The reality is one in which materials keep circulating but largely via the global economy. Wastes and recovered materials flow beyond the EU's borders to be recycled thousands of miles away via Chinese, Indian, Bangladeshi and African firms and labour.

That this is the reality of recycling is indicative of the difficulties of turning environmental values to economic value in particular places. Sustainable futures cannot be produced by ideals alone. Rather, they take finance, labour, economies and markets to make them work. And there is the problem. 'Recycling', as a consumer act of segregating some wastes, captures materials for potential recovery, but those materials require economies – and specifically manufacturers – to value the recovered materials and to then actually recycle them into new products. This valuation is always a matter of purity and contamination. Whilst the physical possibilities might seem endless, the limits to what is possible by way of purity and contamination are defined by economic geography and economics. Our work has shown it is immensely difficult to turn certain recovered materials to sufficient value to cover their cost in certain parts of the world, particularly in Northern Europe. Trying to do so creates some of the toughest occupations and working conditions currently to be found within the EU – and even then the products are not particularly high quality. 'Green' they may be, environmentally contained they may be, but are these jobs emblematic of a sustainable, clean economy? We are not convinced. Rather, these new green recovery-for-recycling jobs have created and reinscribed some of the European labour market's most profound inequalities and divisions. Recycling activity within Europe has been accommodated within the global economy, rather than transforming it. Global recycling networks linked to global patterns of manufacturing have intensified and made for more complex patterns of circulation and exchange. They have commodified wastes by turning them into resources; they show that the trades in commodities and goods is more complicated than overly simplistic notions which portray these as either flows from a producing Global South to consuming Global North or the Global North dumping waste on the poor; and they show that the spatial fix for recycling currently is global. Further social science research along the lines indicated by this chapter is essential in order that environmentally inspired public policy does not have the effect of promoting what is economically challenging or even impossible to achieve, as well as undesirable in terms of the type of jobs that such work creates.

Notes

1. The Waste Hierarchy presents a hierarchy of preferred options for managing wastes, by diverting them from landfill. At its apex is prevention, followed by minimisation. Reuse is next, above recycling and recovery as energy, which includes incineration and newer technologies such as anaerobic digestion. Disposal is at the bottom of the hierarchy and its undesirability is enforced through application of the Landfill Tax. The hierarchy has been used to guide and inform policy, with intervention moving progressively up the hierarchy. However, a key issue is that interventions in the hierarchy can pull in different directions. So, actions to promote recycling or energy recovery, both of which require that wastes be generated at high volumes, can work against waste prevention or minimisation.
2. Examples include the End of Life Vehicles Directive and the WEE Directive for electrical and electronic goods, both of which prohibit movement outside the EU area.
3. Recent examples in the literature which focus on the technical possibilities of using recovered materials to make products include: El-Haway et al. (2010); Nor et al. (2010); Illingworth et al. (2012); Sabai et al. (2013); Schettini et al. (2013). Another strand of literature focuses on improving automated technologies for materials separation and segregation (e.g., Bezati et al., 2011; Lee and Rahimifard, 2012). A rare instance of a technical paper which recognises the importance of economic barriers to reuse is Russell et al. (2010) who identify the cost of insignia removal as a major impediment to the recycling and reuse of corporate clothing.

4. This programme was funded by the Economic and Social Research Council (RES 000 23 0007).
5. Research on textiles was conducted by Lucy Norris; and that on end-of-life ships by Nicky Gregson, Mike Crang and Farid Ahamed. Key references are: Botticello (2012); Norris (2012); Crang et al. (2013) on textiles; and Gregson et al. (2010), Gregson et al. (2012a) and Gregson et al. (2012b) on ship breaking. A short film: *Unravel*, by Meghna Gupta, focuses on textile recycling in Panipat. Additional important work on textiles includes: Tranberg Hansen (2000); Rivoli (2005); Olumide (2012); Brooks (2013). For research on e-waste see Lepawsky and McNabb (2010).
6. The research was conducted by Nicky Gregson, Helen Watkins and Melania Calestani (end-of-life ships); Julie Botticello and Lucy Norris (textiles); and Nicky Gregson, Sara Fuller and Mike Crang (dry recyclables). Further references to research publications appear at the end of the chapter. We draw extensively on this work in this chapter.

Bibliography

Alexander, C. and Reno, J. (eds) (2012). *Economies of Recycling: The Global Transformation of Materials, Values and Social Relations*, London: Zed Books.

Alexander, C., Smaje, C., Timlett, R. and Williams, I. (2009). Improving social technologies for recycling: interfaces, multi-family dwellings and infrastructural deprivation. *Proceedings Institute of Chartered Engineers – Waste and Resource Management*, 162, 15–28.

Asim, M., Batool, S. and Chaudhry, M. (2012). Scavengers and their role in the recycling of waste in Southwestern Lahore. *Resources, Conservation and Recycling*, 58, 152–162.

Bezati, F., Froehlich, D., Massadier, V. and Maris, E. (2011). Addition of X-Ray fluorescent tracers into polymers, new technology for automatic sorting of plastics: proposal for selecting some relevant tracers. *Resources, Conservation and Recycling*, 55, 1214–1221.

Boo, K. (2012). *Behind the Beautiful Forevers: Life Death and Hope in a Mumbai Slum*. London: Portobello Books.

Botticello, J. (2012). Between classification, objectification and perception: processing second-hand cloth for recycling and reuse. *Textile: The Journal of Cloth and Culture*, 10, 164–183.

Brooks, A. (2013). Stretching global production networks: the international second-hand clothing trade. *Geoforum*, 44, 10–22.

Bulkeley, H., Watson, M. and Hudson, R. (2007). Modes of governing municipal waste. *Environment and Planning A*, 39, 2733–2753.

Chappells, H. and Shove, E. (1999). The dustbin: a study of domestic waste, household practices and utility services. *International Planning Studies*, 4, 267–280.

Clapp, J. (2001). *Toxic Exports: The Transfer of Hazardous Wastes from Rich to Poor Countries*, Ithaca: Cornell University Press.

Cooper, T. H (2005). Slower consumption: reflections on product life cycles and the throwaway society. *Journal of Industrial Ecology*, 9, 51–67.

Cooper, T. H (ed.) (2010a). *Longer Lasting Products*, Farnham: Ashgate.

Cooper, T. (2010b). Burying the 'refuse revolution': the rise of controlled tipping in Britain, 1920–1960. *Environment and Planning A*, 42, 1033–1048.

Crang, M., Hughes, A., Gregson, N., Norris, L. and Ahamed, F. (2013). Rethinking governance and value in commodity chains through global recycling networks. *Transactions Institute of British Geographers*, 38, 12–24.

Davies, A. (2008). *The Geographies of Garbage Governance*, Farnham: Ashgate.

Davoudi, S. and Evans, N. (2005). The challenge of governance in regional waste planning. *Environment and Planning C*, 23, 493–517.

Davoudi, S. (2000). Planning for waste management: canging discourses and institutional relationships. *Progress in Planning*, 53, 15–216.

Davoudi, S. (2009). Scalar tensions in the governance of waste: the resilience of state spatial Keynesianism. *Journal of Environmental Planning and Management*, 52, 37–56.

El-Haway, M., Abdul-Jaleel, A. and Al-Otaibi, S. (2010). Recycling crushed concrete fines to produce lime-silica bricks. *Proceedings of the Institute of Chartered Engineers – Waste and Resource Management*, 163, 123–128.

European Environment Agency (EEA) (2009). *Waste Without Borders: Transboundary Shipments of Waste*, Copenhagen: European Environment Agency.

Gregson, N., Crang, M., Ahamed, F., Akter, N. and Ferdous, R. (2010). Following things of rubbish value: end-of-life ships, 'chock-chocky' furniture and the Bangladeshi middle class consumer. *Geoforum*, 41, 846–854.

Gregson, N., Crang, M., Ahamed, F., Akter, N., Ferdous, R., Foisal, S. and Hudson, R. (2012). Territorial agglomeration and industrial symbiosis: Sitakunda-Bhatiary, Bangladesh, as a secondary processing complex. *Economic Geography*, 88, 37–58.

Gregson, N., Metcalfe, A. and Crewe, L. (2007a). Identity, mobility and the throwaway society. *Environment and Planning D: Society and Space*, 25, 682–700.

Gregson, N., Metcalfe, A. and Crewe, L. (2007b). Moving things along: the conduits and practices of divestment. *Transactions Institute of British Geographers*, 32, 187–200.

Gregson, N., Watkins, H. and Calestani, M. (2012). Political markets: recycling, economization and marketization. *Economy and Society*, 13(1): 1–25 (online early: doi 10.1080/03085147.2013)

Illingworth, J., Williams, P. and Rand, B. (2012). Novel activated carbon fibre from biomass fibre waste. *Proceedings of the Institute of Chartered Engineers – Waste and Resource Management*, 165, 123–132.

Institute of Mechanical Engineers (IME) (2013). *Global Food: Waste Not Want Not*, Available at http://www.imeche.org/knowledge/themes/environment/global-food, accessed 10 January 2013.

Kamuk, B. and Hansen, J. A. (2007). Editorial: gobal recycling or waste trafficking in disguise? *Waste Management & Research*, 25(6), 487–488.

Lee, J. and Rahimifard, S. (2012). An air-based automated material recycling system for post-consumer footwear products. *Resources, Conservation and Recycling*, 69, 90–99.

Lepawsky, J. and McNabb, C. (2010). Mapping international flows of electronic waste. *The Canadian Geographer*, 54, 177–195.

McDonald, S. and Oates, C. (2003). Reasons for non-participation in a kerbside recycling scheme. *Resources, Conservation and Recycling*, 39, 369–385.

Minter, A. (2011). The Chinese town that turns your old Christmas tree lights into slippers. Available at http://www.theatlantic.com/international/archive/2011/12, accessed 10 January 2013.

Morley, N., Bartlett, C. and McGill, I. (2009). Maximising reuse and recycling of UK clothing and textiles: final report to the Department for Environment Food and Rural Affairs. London: Oakdene Hollins.

Nor, H., Lim, S. and Ling, T. (2010). Using recycled waste tyres on concrete paving blocks. *Proceedings of the Institute of Chartered Engineers – Waste and Resource Management*, 163, 37–45.

Norris, L. (2012). Economies of moral fibre? Recycling charity clothing into emergency aid blankets. *Journal of Material Culture*, 17, 389–404.

Olumide, A. (2012). The international trade in secondhand clothing: managing information asymmetry between West African and British traders. *Textile*, 10, 184–199.

Packard, V. (1960). *The Waste Makers*, New York: David McKay.

Rivoli, P. (2005). *Travels of a T-Shirt in the Global Economy*, London: Wiley.

Russell, S., Morley, N., Tipper, M., Drivas, I. and Ward, G. (2010). Principles of the recovery and reuse of corporate clothing. *Proceedings of the Institute of Chartered Engineers – Waste and Resource Management*, 163, 165–172.

Sabai, M., Cox, M., Kato, R., Egmond, E. and Lichtenberg, J. (2013). Concrete block production from construction and demolition waste in Tanzania. *Resources, Conservation and Recycling*, 72 9–19.

Schettini, S., Santagata, G., Malinconico, M., Immirzi, B., Scarascia Mignozza, G. and Vox, G. (2013). Recycled wastes of tomato and hemp fibres for biodegradable pots: physic-chemical characteristics and field performance. *Resources, Conservation and Recycling*, 70, 9–19.

Strasser, S. (2000). *Waste and Want: A Social History of Trash*, New York: Owl Books.

Stuart, T. (2009). *Food Waste*, London: Penguin.

Tonglet, M., Phillips, P. and Bates, M. (2004). Determining the drivers for householders pro-environmental behaviour: waste minimisation compared to recycling. *Resources, Conservation and Recycling*, 42, 27–48.

Tranberg Hanson, K. (2000). *Salaula: The World of Second-Hand Clothing and Zambia*, Chicago: University of Chicago Press.

W(aste) R(esources) A(ction) P(rogramme) (2008) *The Food We Waste*, Available at http://www.wrap.org/thefoodwewaste, accessed 13 January 2013.

5
Poverty and Inequality

Rod Hick

Introduction

Despite unprecedented wealth, the problems of poverty and inequality remain important public – and political – concerns. Indeed, the current economic climate perhaps gives them particular relevance. The endurance of poverty and rising levels of inequality impassions 'experts' and 'non-experts' alike.

In this chapter, I discuss some important research in the fields of poverty and inequality. I begin by discussing some of the earliest studies of poverty – the work of Charles Booth and Seebohm Rowntree, conducted in the UK at the end of the nineteenth century. I discuss the evolution of this literature, and the development of Peter Townsend's 'relative deprivation' approach to conceptualising poverty, which continues to provide the over-arching framework for poverty analysis in the UK and Europe, and which can be seen as a response to this earlier literature.

Rather more contemporaneously, in terms of inequality, I discuss Richard Wilkinson and Kate Pickett's, *The Spirit Level,* which was first published in 2009, and which argues that greater levels of inequality in a society result in elevated rates of health and social problems. In the short period since its publication, this book has generated a significant impact and has resulted in a heated debate, and I discuss both the original contribution and the subsequent debate here.

These are, then, two contrasting 'cases' for examining the literature on poverty and inequality. Others might have been selected. But I have chosen to focus on some of the early studies of poverty, on the one hand, and one important, recent debate about the nature of inequality, on the other, because these contrasting cases seem to say something about the nature of social science itself. In particular, these cases represent attempts to systematically analyse poverty and inequality in order to more fully understand the nature of these phenomena, their causes and their consequences.

Analysing poverty

In the UK, poverty analysis is typically traced back to the work of Charles Booth and Seebohm Rowntree. Booth's work examined the extent of poverty in London, initially in 1886 in the city's East End, but with further studies, published in seventeen volumes, covering the whole of the city. Booth sought to 'enumerate the mass of the people of London in classes according to degrees of poverty or comfort and to indicate the conditions of life in each class' (Booth, 1902: 3). To do so, he interviewed School Board Visitors, who themselves had undertaken house-to-house visitations in the relevant areas (Booth, 1887: 327). Booth concluded that 30.7% of the population of London were living in poverty, an alarming finding given London's position as the capital of the richest empire in the world (Bales, 1999).

Seebohm Rowntree's subsequent study of poverty in York in 1899, the first of three such studies, was deeply influenced by Booth's work. In the introduction to the first study, Rowntree noted his desire to evaluate the extent to which 'the general conclusions arrived at by Mr Booth in respect of the metropolis would be found applicable to smaller urban populations' (Rowntree, 1901: xvii). He concluded that the extent of poverty in York was of a very similar order of magnitude – that 27.8% of the population of York were living in poverty which, he argued, was a 'fact of the gravest significance' (1901: 151).

One important difference between these historical analyses and their contemporary quantitative counterparts was the *proximity* of these analysts to people in poverty. The School Board Visitors who Booth interviewed had an intimate knowledge of the streets of London which he was surveying; Rowntree's first poverty survey was a *census* of all working-class households in York. Such studies are invaluable, but they required a degree of time, dedication, and resources which one can scarcely imagine today.

Indeed, contemporary analysis of poverty and inequality takes one of two broad forms – either the analysis adopts a qualitative approach, where the proximity to people in poverty is maintained, but usually at the expense of 'breadth' – by which is meant that the number of people involved in any study falls very far short of Rowntree's census. Or else analysis takes on a quantitative form, where a greater degree of breadth may be achieved, but often at a greater remove from people in poverty themselves. Today's quantitative studies of poverty typically rely on large-scale, secondary datasets or official statistics which were simply not available to Booth and Rowntree.

Perhaps the primary legacy of Charles Booth's work has been his famous poverty maps, in which he depicted the level of poverty on a street-by-street basis. Streets coloured in black were made up mostly of households in Class A, the lowest class, and comprised of '(so-called) labourers, loafers, semi-criminals' (1887: 329). Streets where households were from Class B were depicted in dark blue; these households were very poor and described by Booth as being a people who were 'shiftless, hand-to-mouth, pleasure loving, and always poor' (1887: 329). Streets coloured in light blue represented households from Class C, who were also poor – 'a pitiable class, consisting of struggling, suffering, helpless people' (Booth, 1887: 332). These maps, today housed in the London School of Economics, are an important legacy of this ground-breaking work.

Booth's map can be compared with another important nineteenth century map of London. In 1854, John Snow's 'cholera map' helped to identify the cause of a cholera outbreak in the district of Soho. At the time, it had been assumed that the transmission of cholera could be explained by miasma theory (i.e., that the disease was airborne). Snow used a street map of the district and plotted the location of the cholera deaths, demonstrating that what most had in common was proximity to a particular water pump in Broad Street. The handle of the water pump was removed, and cholera deaths began to fall over the following days. Snow's work, and his map, played a role in the advancemment of what was then known as germ theory.

The distinction between these two maps highlights something important about the analysis of poverty and inequality – these are, at least partially, *observable* phenomena. And this, in turn, matters because it suggests that the findings which emerge from academic poverty analysis might well be consistent with the perspectives of journalists, think tanks, and indeed the general public.

While the scale and suggested causes of poverty and inequality in Booth's study may have been alarming, his focus on East London (as opposed to, say, West London) would not have been. The social sciences seek to bring a kind of *rigour* to the study of poverty and inequality, but it cannot be assumed that people's own understandings about poverty and inequality are somehow deficient.

A short stroll around any major 'developed' city tells its own story in terms of street homelessness – perhaps the most visible, if not prevalent, form of poverty and deprivation co-existing, often cheek-by-jowl with unprecedented affluence. Similarly, any visitor to, for example, Hong Kong, or Rio de Janeiro, or Manila will observe the alarming levels of inequality in those cities. One simply does not need a social scientist to confirm astounding poverty and inequality in the world in which we live in the way that, say, we may need a scientist to detect the presence of a particular gas in the atmosphere. There is much about poverty and inequality which seems readily accessible to the 'non-expert'.

The studies of Booth and Rowntree, however, sought not only to investigate the conditions and prevalence of poverty; they also sought to understand the *causes* of poverty – including whether poverty was, in any sense, because of people's own behaviour and whether they could have avoided falling into poverty. Rowntree had identified households as being in poverty where they were living in 'obvious want and squalor'; in addition, however, he divided this group into households living in 'primary' poverty and others living in 'secondary' poverty. Primary poverty focused on cases where household resources fell below the 'minimum necessary expenditure for the maintenance of merely physical health' (Rowntree, 1901: 118). 'Secondary' poverty, on the other hand, referred to a circumstance where a household's earnings 'would be sufficient for the maintenance of merely physical efficiency were it not that some portion of it is absorbed by other expenditure, *either useful or wasteful*' (1901: 118, emphasis added).

For example, Rowntree's Class A (households with incomes below 18 shillings a week for a moderate family) were, he claimed, unable to avoid the experience of poverty; the chief reasons for poverty in this class being the 'continued lack of work, or the death or illness of the chief wage-earner' (1901: 74). The vignettes provided by Rowntree illustrate the position of particular households – this household falling into Class A:

> Husband in asylum. Four rooms. Five children. Parish Relief. Very sad case. Five children under thirteen. Clean and respectable, but much poverty. Woman would like work. The house shares one closet with another house, and one water-tap with three other houses. Rent 3s. 9d. (1901: 63)

In contrast, in households in the better-off Class D, 'there is, practically speaking, no poverty…except such as is caused by drink, gambling, or other wasteful expenditure' (1901: 104). This reflects a general concern to understand the extent to which people were themselves 'responsible' for their own poverty and, the role played by drink, in particular, which emerges as an important theme in these early studies.

It may seem hard to believe today that there were genuine questions about whether there was any 'real' poverty in nineteenth century Britain, but these questions were indeed genuinely asked – as they are today. And – as is the case today – these questions required a rigorous response. There is an important parallel between the questions which these early studies sought to address, and contemporary concerns about whether poverty is caused by unemployment, on the one hand, or because of people's own behaviours, on the other.

Need and poverty

Unlike inequality, which can be analysed using more or less arbitrary proportions of a distribution,[1] the study of poverty requires a cut-off; a poverty line which conveys some *meaning*. Often, this meaning is based on the idea of human *need*. The idea that people have needs which can be distinguished from more frivolous desires or 'wants' has widespread and, indeed, intuitive appeal. But teasing out exactly what it is that people need, and distinguishing these from mere wants, is a challenging endeavour.

In identifying a budget in order to estimate his primary poverty line, Rowntree considered three types of expenditure: food, housing costs, and 'household sundries (such as clothing, light, fuel, etc.)' (1901: 119). In estimating the cost of a food budget, for example, Rowntree sought to identify the minimum number of calories that people required each day, and then a diet which represented the cheapest way in which this calorific intake could be achieved. In selecting the particular diet on which he would base his poverty line, Rowntree turned to those foods provided to able-bodied paupers in workhouses, 'as the object in these institutions is to provide a diet containing the necessary nutrients at the lowest cost compatible with a certain amount of variety' (1901: 129–130). This approach was deliberately minimalistic: one of Rowntree's central concerns was to ensure that no one could claim that he had set his primary poverty threshold *too high* (e.g., 1901: 129, 1941: 28).

Rowntree noted that successfully identifying the cheapest way to achieve the necessary nutrients would be considerably demanding of households:

> It must be remembered that at present the poor do not possess the knowledge which would enable them to select a diet that is at once

as nutritious and as economical as that which is here adopted as the standard. Moreover, the adoption of such a diet would require considerable changes in established customs, and many prejudices would have to be uprooted. (Rowntree, 1901: 137)

It was on the basis of this extremely restrictive poverty standard that Rowntree found that 9.91% of the population of York were, in 1899, living in primary poverty. Rowntree claimed that the causes of primary poverty were, in descending order of importance, low wages (but in employment), largeness of family, death of chief wage earner, illness or old age of chief wage earner, irregularity of work, chief wage earner out of work (1901: 154). From this he concluded that 'the wages paid for unskilled labour in York are insufficient to provide food, shelter and clothing adequate to maintain a family of moderate size in a state of bare physical efficiency' (1901: 166).

In Rowntree's second and third studies of poverty in York, these poverty standards were uprated. His poverty standard for the second, 1936 survey was devised using five budget headings: food; clothing; fuel and light; household sundries, and personal sundries (1941: 28). These personal sundries included amounts for unemployment insurance, a daily newspaper, a wireless radio and 3s. 4d. set aside for 'all else', suggested to account for 'beer, tobacco, presents, holidays, books, travelling, etc.' (1941: 28). This budget was uprated again in his third, and final, survey in 1950 (Rowntree and Lavers, 1951). By the time of the second, 1936 survey, primary poverty was found to have fallen to 3.9% of the total population, while the proportion of the population in the worst-off classes, A and B, fell from 17.7% in 1936 to 1.7% of the population of York in 1950 (Rowntree and Lavers, 1951: 30–31), suggesting that poverty had, by and large, been eliminated.

The work which has had perhaps the greatest influence on understanding poverty today, however, has been that of the sociologist Peter Townsend. In outlining his own 'relative deprivation' approach to understanding poverty, Townsend juxtaposed his work against the earlier studies of Rowntree. Townsend claimed that although Rowntree had uprated his poverty standard over time, he had failed to provide a convincing account for *why* this uprating had occurred (see Townsend, 1970: 13).

Townsend argued that there could be no successful attempt to identify 'absolute' needs relating to, for example, 'merely physical efficiency', because even supposedly absolute needs were influenced by social customs and norms. For example, people's eating habits were influenced not only by nutrition, but also occupational and leisure patterns. Townsend offered the example of a cup of tea: nutritionally worthless, but which played an important social role. If custom dictated that visitors to a house would expect to be offered a cup of tea, then this would impose resource demands on households to meet this need. To focus solely on minimum calorific requirements would be to ignore how people actually lived.

In his classic *Poverty in the United Kingdom*, Townsend argued that Rowntree had failed to provide an adequate approach to understanding need. First, he argued that Rowntree had underestimated the cost of necessities other than food; second, that even the food budgets provided calorific minima which were 'very broad averages not varied by age and family composition, still less by occupation and activity outside work' (Townsend, 1979: 34); third, that the foods selected took no account of customary diets at the time.

In contrast, Townsend argued that poverty analysis must consider the full range of human needs, and that these could only be considered relative to the societies in which people lived because people's needs were socially determined. The central advance of the Townsendian conceptualisation, then, was to argue that a poverty standard must evolve over time in line with changes in social customs and expectations – to argue that poverty was *relative*.

Rather than an expert prescribing minimum calorific requirements relating to physical needs, what was required, Townsend believed, was to undertake an extensive survey of the 'style of living' in order to understand the social customs which people were expected to uphold and the activities they were expected to participate in. In his own research, Townsend pioneered the use of indicators of material deprivation in poverty analysis. Respondents were asked whether they did not have fresh meat as many as four days a week, whether the household did not have a refrigerator, and whether the members of a household did not have a week's holiday away from home in the previous year (Townsend, 1979: 250), among others. Townsend hypothesised that respondents' deprivation scores could be used to identify a threshold in the income distribution, below which people's withdrawal from sharing in widespread customs and activities would escalate disproportionately. He plotted his respondents' deprivation scores across the income distribution and claimed to find suggestive evidence of such a threshold.

The Townsendian conceptualisation continues to provide the underpinning for poverty analysis to this day in the UK and the EU. Indeed, the University of Bristol, in tandem with a team of researchers at other

Universities, is currently undertaking a major analysis of poverty in the UK (www.poverty.ac.uk). As part of this research, they are seeking to identify 'which items and activities...should be seen as necessities for living in the United Kingdom today'. Their focus is on what survey respondents themselves understand to be the commodities and activities which everybody should be able to afford in today's society; items which no one should have to do without. This systematic attempt to use deprivation indicators as a means of identifying what the public themselves understand to be necessities is indicative both of the rigorous nature of social scientific research on poverty and of the enduring legacy of the Townsendian perspective, first advanced some sixty years ago and which continues to inspire poverty analysis in the twenty-first century.

Analysing inequality

In addition to poverty, there has, in recent years, been an increased focus on *inequality* within Great Britain and, indeed, beyond, perhaps most visibly associated with the Occupy movement, which has sought to challenge inequality, focusing in particular on the concentration of income and wealth on the top 1% of distribution.

In the present chapter, we discuss the publication and subsequent debate surrounding Richard Wilkinson and Kate Pickett's *The Spirit Level* (2009). Wilkinson and Pickett argue that, for much of human history, achieving human progress was inextricably linked to society becoming wealthier. However, they claim that the developed world is faced with a new problem because 'economic growth, for so long the engine of progress, has, in the rich countries, largely finished its work' (2009: 5).

Wilkinson and Pickett display an association between a measure of average wealth (national income per person) and, first, life expectancy, and second, happiness for a large number of countries, drawing on data from the United Nations. These associations show, they claim, that while there is a relationship between wealth and life expectancy and happiness for poorer countries, beyond a certain point further increases in wealth are no longer associated with additional rises in life expectancy and happiness. This is not, they suggest, because we have reached the limits of human progress – life expectancy, for example, continues to increase – but rather that social progress is no longer dependent on average wealth (2009: 6).

Wilkinson and Pickett present their subsequent thesis using two sets of data, analysing cross-national data from 23 rich nations, and from the fifty states of the US. These two datasets are used to explore the relationship between average wealth, inequality and an index of health and social problems. Most of the empirical analysis which is presented in *The Spirit Level* consists of a series of scatter plots which demonstrate the relationship between economic growth or income inequality and this index. The index is comprised of 'levels of trust, mental illness (including drug and alcohol addiction), life expectancy and infant mortality, obesity, children's educational performance, teenage births, homicides, imprisonment rates, social mobility' (2009: 19). Wilkinson and Pickett show, using both cross-national and US data, that there is only a weak relationship between average income and the index of these health and social problems (2009: 21). This contrasts with the very clear and, indeed, strong relationship between income inequality and the index of health and social problems, using both datasets.

Having established this distinction for the index as a whole, the book then turns to analysing the relationship between inequality and the individual components of the index of health and social problems. The reader is shown that more unequal societies, and more unequal US states, have lower levels of trust in others, greater rates of mental ill-health, lower life expectancy, greater infant mortality, more obesity, poorer educational performance, a greater rate of teenage pregnancies, and murders, incarcerations, and lower social mobility. Based on the evidence presented, life does seem very much worse in unequal societies. Wilkinson and Pickett claim that the evidence shows that 'almost all social problems which are more common at the bottom of the social ladder are more common in more unequal societies' (2009: 18).

The reason for these findings is, they suggest, that economic inequality is itself a marker of status differences, and it is these status differences which 'get under the skin' in the form of 'status anxiety'. Greater inequality, they suggest, emphasises the importance of social status (2009: 43), and it is the psychosocial effects of inequality which Wilkinson and Pickett claim are the explanatory mechanism by which material inequalities – and not absolute material wealth – result in negative health and social outcomes for a society.

They argue that:

> The problems in rich countries are not caused by the society being rich enough (or even by being too rich) but by the scale of material differences between people within each society being too big. What matters is where we stand in relation to others in society. (2009: 25)

Theirs is a big finding, and, if correct, would contain significant policy implications. They note that ours is 'the first generation to have to find new answers to the questions of how we can make further improvements to the real quality of human life' (2009: 11), and while they stress that there can be different routes to arriving at a more equal society (2009: 243), it seems that a considerably greater degree of redistribution from the rich to the poor would be a necessary policy response, at least in a relatively inegalitarian society such as the UK.

Response to *The Spirit Level*

Such a big finding has, inevitably one might say, generated significant debate, much of which has been focused on methodological issues. Two works, by Christopher Snowdon (2010) and Peter Saunders (2010), are of particular interest, since both present a reanalysis of Wilkinson and Pickett's work in order to examine the *The Spirit Level*'s claims.

The first issue raised in these critiques is Wilkinson and Pickett's selection of countries, and the influence of this selection on the results that follow. Wilkinson and Pickett (2009: 275) claimed to start with a list of the richest 50 countries in the world and then exclude any country with a population of below three million because they wanted to exclude tax havens, and others for whom income inequality data were not available. Saunders (2010: 21) argues that a lower threshold of 1 million serves to exclude tax havens such as Monaco and the Cayman Islands, but results in the inclusion of six additional countries – Slovenia, Trinidad and Tobago, Estonia, Latvia, Gabon and Botswana – not typically understood as being tax havens. Snowdon, too, finds something amiss with the selection criteria which, he claims, 'only serves to exclude Slovenia from the analysis' (2010: 13). He claims that 'puzzlingly, the Republic of Korea and Hong Kong were left out entirely, despite their wealth' (2010: 13), and asks why Czech Republic, Korea and Slovenia have been excluded given that they are all wealthier than Portugal, which Wilkinson and Pickett had included (Snowdon, 2010: 153).

Furthermore, Saunders claims that the 2004 Human Development Report (one of Wilkinson and Pickett's sources) contains income data for far more countries than they acknowledge. Saunders finds 44 nations which conform to his criteria of being amongst the 50 richest countries, with populations above 1 million and data on income inequality (this time using the 2009 Human Development Report, see 2010: 21).

Snowdon (2010) claims that if unnecessarily excluded countries are reinserted, some of the relationships – for example, between inequality and obesity – disappear. Such a claim is of course damaging for Wilkinson and Pickett's suggestion that there is a general pattern in the relationship between inequality and health and social problems. Similarly, in his 2009 review in the *Financial Times*, John Kay noted that the analysis presented in *The Spirit Level* was based 'mostly on a series of scatter diagrams' which appear 'dominated by a few outliers'. If one removed the US (a known 'poor' performer) and Japan and the Scandinavian countries (known 'good' performers), many of the patterns would, Kay argues, be random.

Saunders (2010) also points to the reliance on outliers to find a clear international trend, and offers two instructive examples. First, he claims that the association between income inequality and homicide is attributable solely to the US, which, he posits, might have a higher homicide rate due to its gun laws rather than because of its inequality. Secondly, he claims that the association between inequality and life expectancy is attributable to the inclusion of Japan, whose long life expectancy, he argues, might be attributable to their diet, or genes, or both, rather than their low income inequality (2010: 7). Having reanalysed 20 of Wilkinson and Pickett's empirical claims, Saunders argues that only one commands unambiguous support, a further five are found to have ambiguous support, while 14 are, he claims, 'either spurious or invalid' (2010: 6).

In addition, some authors have questioned Wilkinson and Pickett's claims of *causality*. In her review of the evidence, Rowlingson (2011: 5) argues that while many health and social problems have a clear social gradient (i.e., that they are experienced more by the poor than the rich), 'there is less agreement about whether or not there is a causal relationship'. John Goldthorpe (2010: 737) has criticised Wilkinson and Pickett for their 'inadequate, one-dimensional understanding of social stratification' which, he claims, 'leads to major problems' when seeking to understand *why* inequality leads to these health and social problems. Goldthorpe questions the idea that it is income inequality *as a marker of social status* which generates poor health and social outcomes. In doing so, he is questioning the psychosocial process which Wilkinson and Pickett offer as the key to understanding why these problems are more prevalent in unequal societies. Goldthorpe points to the case of Japan which, while appearing at the 'good' end of the scatter plots (i.e., relatively low levels of health and social problems, and low levels of income inequality) is nonetheless a country where status differences are extremely pronounced (2010: 738) – it is just that these status differences do not manifest themselves in substantial income inequalities. So Goldthorpe claims that Wilkinson and Pickett's explanatory mechanism seems to fail to account for the case of Japan and, more generally, that economic advantage and disadvantage display a far more variable association with status differences than they assume.

Snowdon (2010: 61) objects to the persistent and exclusive focus on inequality as driving the trends which are displayed – arguing that Wilkinson and Pickett 'seem unable to look beyond inequality for an explanation' for the relationships they observe. This leads, he argues, to overlooking many particularities about the nations that are analysed:

> It takes a particularly blinkered view of the world to portray Sweden, a country which has not fought a war since 1814, as being fundamentally the same as Israel, except in its distribution of wealth … Or to imagine that a culturally homogenous, traditional Asian country like Japan can only be distinguished from the United States by reference to the gap between the richest and poorest 20% of the population. (Snowdon, 2010: 136)

Saunders, too, argues that Wilkinson and Pickett ignore cultural or historical explanations for their trends, and that they do not sufficiently investigate the importance of third variables, which explain the association between inequality and their index of health and social problems. For example, in his cross-national analysis, Saunders claims that while inequality is related to trust, 'the effect of GDP is greater' (2010: 44). In his analysis of US states, he argues that the proportion of African–Americans is a stronger determinant of some of the social indicators in question (e.g., homicide, infant mortality, life expectancy, teenage births, imprisonment) than inequality *per se*, drawing on multivariate analysis. This line of criticism argues that Wilkinson and Pickett have failed to sufficiently examine the many other explanations which might account for the trends they present.

Finally, Saunders (2010: 8) argues that Wilkinson and Pickett ignore social problems which are *more prevalent* in equal societies – suicide rates, HIV infection rates, alcohol consumption and divorce rates are higher, and fertility is lower in more equal societies, something which is noted by Snowdon, too.

As if to inject some levity into their works, both Snowdon (2010) and Saunders (2010) present more light-hearted analyses in their concluding arguments. Snowdon provides a scatter plot of the relationship between distance from each country's capital city to the North Pole and educational achievement – a significant, negative relationship is demonstrated. Being in a country close to the North Pole

(e.g., Finland or Sweden) is associated with better educational performance, while living far from the North Pole (e.g., Israel) is associated
with lower educational performance. This serves to emphasise the fact
that correlations between variables cannot be taken to imply a causal
relationship.

Drawing on the indicators which are more common in equal societies, Saunders (2010) comprises what he calls a 'Social Misery Index',
comprised of racist bigotry (minding whether a neighbour is of a different
race), suicide rate, divorce rate, fertility rate, alcohol consumption and
HIV infection rate (2010: 106). He plots this index against inequality
and finds that there is less Social Misery in more unequal societies. Of
course, he isn't seeking to put this forward as an alternative hypothesis;
rather, to show that a wider selection of social ills does not conform to
Wilkinson and Pickett's hypothesis.

The central point about this debate, however, whatever the position
one takes within it, is that it is *precisely* the type of debate one might
expect from a social science. Arguing about whether the right cases have
been selected, whether the results are sensitive to the selection of particular cases or are generalisable to a broader set, whether outliers have a
significant bearing on the results, whether there is sufficient evidence
of the hypothesised causal pathways, and whether other, competing
hypotheses have received sufficient attention is the very business of
social science. It is by asking these questions that we work towards the
rigour which is the goal of the social sciences.

Concluding discussion

Poverty and inequality remain major problems in this second decade
of the twenty-first century. These problems generate significant debate,
both between the academics who study them and amongst the wider
public, from whom they often provoke strong views. As Wilkinson and
Pickett (2009: x) note:

> Controversies in the natural science are usually confined to the
> experts: most people do not have strong views on rival theories in
> particle physics. But they do have views on how society works.

There is much about studying poverty and inequality, we have argued,
which is not dependent on 'experts'. Both are at least partially observable phenomena, and this creates an accessibility which opens up the
analysis of poverty and inequality, not only to a wide readership, but

also, potentially, to a wide *authorship* – journalists, think tanks, and the general public have an important role to play, too.

If the social sciences can claim any particular expertise or distinctiveness in this area, it is surely in terms of the *rigour* to which they aspire through an appeal to evidence, scepticism, examination and replication. This systematic, rigorous approach can be seen in the studies discussed here – in the attempts to identify poverty in nineteenth century London on a street-by-street basis; in the attempt to identify a primary poverty line which allows households to achieve 'merely physical efficiency'; and in more recent attempts to examine *which* commodities and activities are understood by people in Britain today to be necessities. It can also be seen in the attempt, first, to analyse the relationship between inequality and health and social problems, and subsequently, by others who criticise, examine and – crucially – replicate the initial analysis. Often the rigour which the social sciences seek remains an aspiration; it is not always translated into reality. But it is by striving towards this rigour that the social sciences can contribute, in a unique and important way, to understanding the nature and relevance of problems such as poverty and inequality, as they exist in the twenty-first century.

Note

1. For example, by analysing the multiple of income held by the top 20% of income earners compared to the bottom 20%, or 80:20 ratio.

Bibliography

Bales, K. (1999). Popular reactions to sociological research: the case of Charles Booth. *Sociology*, 33(1), 153–168.

Booth, C. (1887). The inhabitants of Tower Hamlets (School Board Division), their conditions and occupations. *Journal of the Royal Statistical Society*, 50(2), 326–401.

Booth, C. (1902). *Life and Labour of the People in London: Religious Influences*, London: Macmillan & Co.

Goldthorpe, J. H. (2010). Analysing social inequality: a critique of two recent contributions from economics and epidemiology. *European Sociological Review*, 26(6), 731–744.

Kay, J. (2009). The Spirit Level: review by John Kay. *Financial Times*, 23 March 2009.

Rowlingson, K. (2011). *Does Income Inequality Cause Health and Social Problems*. York: Joseph Rowntree Foundation.

Rowntree, B. S. (1901). *Poverty: A Study of Town Life*. London: Thomas Nelson & Sons.

Rowntree, B. S. (1941). *Poverty and Progress: A Second Social Survey of York*. London: Longmans, Green and Co.

Rowntree, B. S. and Lavers, G. R. (1951). *Poverty and the Welfare State: A Third Social Survey of York Dealing Only With Economic Questions*, London: Longmans, Green and Co.

Saunders, P. (2010). *Beware False Prophets: Equality, the Good Society and The Spirit Level*, London: Policy Exchange.

Snowdon, C. (2010). *The Spirit Level Delusion: Fact-checking the Left's New Theory of Everything*, North Yorkshire: Little Dice.

Townsend, P. (ed.) (1970). *The Concept of Poverty*. London: Heinemann.

Townsend, P. (1979). *Poverty in the United Kingdom: A Survey of Household Resources and Standards of Living*, Middlesex: Penguin.

Wilkinson, R. and Pickett, K. (2009). *The Spirit Level: Why Equality is Better for Everyone*, London: Penguin.

6
The Economy, Financial Stability and Sustainable Growth

Jonathan Michie

Introduction

There have been many seminal studies of the way the economy operates, and of how economic activity relates to other areas of life studied by the social sciences, most obviously society and politics. In 1776, *The Wealth of Nations* by Adam Smith became the most comprehensive analysis and description up to that time of the way the economy functions – and of the direct implications of this for society, for example regarding the division of labour at work, and the division of income and wealth between social classes. *The Wealth of Nations* remains widely cited, perhaps most commonly as regards the 'invisible hand' of the market. Just under a hundred years later, in 1867, Marx published Volume 1 of *Capital* which sought to explain the 'laws of motion' of capital – that is, of the capitalist economy and of the individual firms that drive it. The failure of the 'invisible hand' to bring together idle resources and unmet needs – creating the sort of economic slump that Marx had analysed – led to another great work of economics just under seventy years later in the form of Keynes's *General Theory of Employment, Interest and Money* in 1936. While Adam Smith enjoyed a renaissance during the decades of privatisation, deregulation, demutualisation and monetisation from the Thatcher/Reagan era of the 1980s onwards, with the Adam Smith Society gaining prominence in the UK, the works of Marx and Keynes have regained some prominence since the global financial crisis of 2007–2008, leading as it did to the first global recession since the 1930s, in 2009.

This chapter considers the contribution of economics and economists in analysing and explaining developments in the economy, and to society more generally, given the interconnections and interdependencies between economic, social and political life.

It is worth reflecting, though, firstly that Smith, Marx and Keynes all surveyed the whole field of economic activity, and of economics as it existed as a discipline in their times, in contrast to today's 'division of labour' within economics itself, not only between macroeconomics and microeconomics, but to far more specialised sub-disciplines of monetary economics, labour economics, industrial economics, public economics, international economics, and so forth. The discipline is also categorised between 'theory' and 'empirical', with the former having sub-categories such as game theory, and the latter having sub-categories such as econometrics.

Secondly, Smith, Marx and Keynes all commented far beyond the realms of the economy alone, considering not only the implications of and for society, but also the political, historical and institutional context of the economy and society they were analysing. They thus had what today would be considered a rather heterodox approach to the subject as opposed to today's neoclassical orthodoxy, along with rather interdisciplinary and multidisciplinary instincts when it came to considering the big questions of the day. It may be that there are lessons to be learned from such authors as regards these issues of methods and approach, as well as regards the more specific matters on which they are usually quoted – regarding the function and functioning of markets, and what happens when such functioning appears to break down, resulting in crises, recession and unemployment.

First, though, this chapter considers what economics does have to say about the 'big' questions of economic growth, globalisation, crises, and corporate diversity and the functioning of markets, before going on to consider how economics and social sciences more generally might develop in the future to better understand and influence the development of the economy and the economic aspects of society.

Measuring economic growth

Adam Smith is well known for pointing to the 'extent of the market' as being important in creating the conditions for economies to grow. He argued that this develops alongside the division of labour within the workplace – in Adam Smith's case, the pin factory. As the workplace expands, it allows a greater division of labour which in turn enables workers to specialise and become more productive, increasing productivity which brings down prices which boosts sales, thus extending the market which in turn enables firms and workplaces to expand, enabling a still greater division of labour.

Adam Smith also famously described the way market exchange can enable the butcher, brewer and baker to respond to the demand for their products through their own (economic) self-interest, but thereby making everyone better off. It was this argument in particular which today tends to have Adam Smith held up in support of 'the market', as opposed to relying on state intervention, although Smith himself did appreciate and discuss the need for government activity alongside the 'market'.

The size of the economy is measured by the 'value' of all goods and services sold or provided. These may be provided free of charge (e.g., by government), but a monetary value can still be attached to all such goods or services, so that the size of the economy can be estimated. Economic growth is then the increase in the size of the economy from one year to the next. There has long been debate and discussion over how important economic growth is compared to broader concepts such as the quality of life, and today this tends to involve questions of environmental sustainability. There are also more technical questions as to whether current measures of economic growth adequately capture changes in the nature of goods and services, the role of intellectual capital and ideas, and the contribution of natural resources. Thus, for example, the former IBM executive, Irving Wladawsky-Berger argues that:

> GDP is essentially a measure of production. While suitable when economies were dominated by the production of physical goods, GDP does not adequately capture the growing share of services and the production of increasingly complex solutions that characterise advanced economies. Nor does it reflect important economic activity beyond production, such as income, consumption and living standards. (Cited in Kaminska, 2013)

It is certainly true that measures of economic growth could and should be improved: firstly, in the more narrowly technical sense of, for example, taking care to use appropriate measures of services, which in many cases may be provided by the public sector at no charge or at least at a price less than the cost of providing them, and taking proper account of natural resources and whether they are being depleted or not; and secondly in the broader sense of ensuring that economic growth is sustainable in environmental and other ways, and also that other factors such as people's wellbeing and happiness are included as policy objectives alongside the narrower focus on economic growth. On the need to reform the way we measure economic growth in order to include the contribution of natural resources, and whether they are being depleted over time, see the first report from the Natural Capital Committee (2013). The idea of attempting to measure happiness has been discussed in the UK by, most notably, Andrew Oswald and Richard Layard. In 2008, French President Sarkozy commissioned Joseph Stiglitz, Amartya Sen and Jean-Paul Fitoussi to consider the limitations of GDP as a measure of economic prosperity and progress. Their Report (Stiglitz et al., 2009) acknowledged all the points made above, and concluded that there does indeed need to be a shift away from just measuring economic outputs towards a greater consideration of people's wellbeing; that more prominence needs to be given to the distribution of consumption, income and wealth; and that environmental sustainability needs to be paid particular attention, within the context of ensuring economic sustainability more generally.

Economic booms and slumps

John Maynard Keynes was certainly one of the greatest economic writers to date, as well as having been active as a policy adviser (not to mention as a College Bursar and in other roles). He first came to the public's attention with his critique of the policies being pursued by the First World War's victorious powers in imposing reparations payments on Germany. His *Economic Consequences of the Peace* (Keynes, 1919) predicted the catastrophic effects that this policy would have on the German economy, and warned against such a short-sighted and self-defeating approach, sadly to no avail. Next he warned against Winston Churchill's equally foolhardy policy of returning Sterling to the Gold Standard at an uncompetitive exchange rate. His *Economic Consequences of Mr Churchill* warned that real world economies simply don't work like the economic textbook models, which was the incorrect assumption on which the Chancellor of the Exchequer's approach was based (Keynes, 1925). Again, Churchill was proved wrong, Keynes's advice was proved right, and the Gold Standard collapsed, but only after having caused serious economic damage.

His *General Theory* (Keynes, 1936) argued that the orthodox economics of his day – as advocated by the Treasury, then and now – was simply wrong in its assumptions as to how the economy worked. The Treasury view was that monetary policy could ensure the economy recovered, with wages adjusting as necessary to ensure a return to full employment. Keynes pointed out, again, that wages don't and won't adjust in the real world as they do in textbooks. He pointed out that attempts to cut wages in monetary terms would be resisted for all sorts of reasons, including the uncertainty amongst those whose wages were to be cut as to what this would mean in terms of comparative earnings compared to others in society. This was one of the warnings he had already issued in his *Economic Consequences of Mr Churchill*, when Churchill as Exchequer had nonetheless returned to the Gold Standard with the aim of making Sterling competitive by cutting wages, which when applied to the mine-workers resulted in Britain's first and, to date, only General Strike, in 1926. In the *General Theory*, Keynes also pointed out that relying on monetary policy alone may be like pushing on a piece of string – if no one is pulling at the other end, the exercise may prove futile, and the same is true if companies don't want to borrow money, for example because they have insufficient confidence that consumers would want to buy the extra goods that the loan would enable them to produce. One reason the industrialists may lack confidence that consumer demand would be sufficient might be if Government was at the same time cutting the wages of those potential consumers.

Instead, Keynes argued, recessions and unemployment are caused by lack of aggregate demand, and if there is not the required demand forthcoming from overseas, which could lead to export-led growth, nor from consumers or, therefore, from companies wishing to invest to meet expected demand from either of those sources, then there is only one other possible source of demand, and that is from Government. In such situations, active fiscal policy is required to boost demand, get the economy moving, create jobs and hence increase consumer demand which in turn will encourage businesses to invest in order to meet the anticipated rise in demand for goods and services. While the role of demand is what Keynes is best known for, he also warned about the behaviour of unregulated markets, with the danger of herd behaviour leading to stock-exchange bubbles. He thus argued for economic regulation at national and international levels to enable markets to operate productively, avoiding unsustainable bubbles and the ensuing crashes and recessions.

The 2007–2008 financial crisis

In the 1980s, though, Keynesian economics was largely overthrown by the more free-market approaches of monetarism and Milton Friedman, and of privatisation and deregulation. For almost thirty years, we saw 'capitalism unleashed' as the late Andrew Glyn depicted it (Glyn, 2007). The backdrop to this was the economic problems of the 1970s when

Keynesian policies appeared to be leading to unsustainable inflation. Milton Friedman's monetarism was presented as an explanation of this inflation – as being caused by excess growth in the money supply. And since unsustainable inflation could not be accepted, this alternative approach, of targeting the money supply in order to bring inflation under control, was argued to be the only alternative.

It should be stressed that this was just one school of thought within economics, and it was consistently critiqued at the time, most notably by Lord Kaldor in the UK and J. K. Galbraith in the US. Other critics included Meghnad Desai (see Desai, 1981) and David Hendry (see Hendry, 1980; Hendry and Ericsson, 1983).

Within the UK, it might be said that the social sciences in general were out of favour with the Thatcher governments of the 1980s (indeed, the Prime Minister declared that there was 'no such thing as society'). The word 'science' was removed from the Social Sciences Research Council, with Keith Joseph (who was the chief *laissez-faire* ideologue in Government along with Thatcher) renaming it the Economic & Social Research Council – thus both removing 'science' and boosting the importance of economics. And the Open University came under attack from Joseph when one of their economic texts described Britain as a capitalist economy.

Economists and other social scientists did of course remain active in analysing the effects of this new policy agenda (e.g., the various contributors to Michie, 1992, including, for example, the implications of unemployment for wellbeing, by Burchell, 1992). But while the nature of governments of course varied over the 1980s, 1990s and into the twenty-first century, both in the UK and internationally, in broad terms the privatisation and deregulation persisted, and the inequality of income and wealth increased, with a redistribution of income away from wages and salaries towards profits and executive remuneration. The push for continued deregulation therefore continued, being advocated by those who were benefiting materially from such policies. Thus President Clinton repealed the Glass-Steagall Act which had been introduced in response to the 1929 Wall Street Crash, to try to prevent a recurrence; this deregulation enabled high street banks to move into more speculative investment bank activities. In the UK, demutualisation led to most of the large building societies, which had been owned by their customers, being converted to shareholder-owned banks, all of which subsequently failed (such as Northern Rock) or were taken over by large shareholder-owned banks, thus reducing the corporate diversity of the financial services sector.

These developments created – or recreated, following the 1929 Wall Street Crash – the conditions for a spectacular collapse. And when it hit, the impact was global, thanks again to the deregulatory policies that had removed what buffers Keynes and others had managed to have put in place during the negotiations at Bretton Woods over the shape of the post–Second World War regulatory architecture.

Globalisation

One of the first acts of the Thatcher Government in 1979 had been to abolish exchange controls. Up until this point, the transfer of currencies internationally was regulated, other than the US dollar that was tied to gold. This all changed from 1979, with other countries following the UK's lead, and a huge rise in capital and currency movements developing globally. This triggered what has been generally referred to as an era of globalisation, with a huge increase in the global movement of money, but also of trade, investments, and other economic activity, and also social, political, cultural and technological developments which all chimed with the idea of a new 'global village', where the 'world is flat' (Friedman, 2005).

The extent and nature of this era of globalisation has been usefully analysed by a range of social scientists and economists. Following his tenure as Chief Economist at the World Bank, Stiglitz published a sceptical analysis of the form and nature of the free-market variant of globalisation policies that had been advocated by the 'Washington consensus' of the International Monetary Fund and the World Bank (Stiglitz, 2002). Ha-Joon Chang questioned whether it was right to say that free-market approaches should be adopted by all, when the currently rich countries had only achieved that status through active government-sponsored industrial policies, generally protected behind exchange controls and import controls (Chang, 2002).

But the *laissez-faire* model of globalisation continued. Keynes had warned of the risks of such an approach at both the national and international levels. 'When the capital development of a country becomes a by-product of a casino, the job is likely to be ill-done', he had concluded in the *General Theory*, having described how stock-exchange behaviour could come to be driven not by economic fundamentals but by the expectations of traders as to the likely trades of the other participants, who in turn were betting on the expected behaviour of others. Expectations can thus become self-fulfilling prophesies. While Keynes's stress on the importance of expectations, risk and uncertainty in the way economies operate were rather underplayed in the decades leading up to the 2007–2008 global financial crisis, the question of risk was sometimes referred to, but only within the damagingly complacent context of the claim that risks had been dealt with, it was suggested, as these had been insured against, including via innovative new financial products.

When the 2007–2008 global financial crisis hit, this argument was shown to have been false, a fallacy of composition, since if one insured contract is called in, it can be paid out and the argument holds, but if they are all called in – or even just a large number – then the insurer cannot pay and instead goes bankrupt, and the bubble bursts just as Keynes had warned. Thus, it should be stressed, while the results of 2007–2008 have been damaging and costly across the world, there is nothing that has happened that had not been warned against by social scientists – from Keynes onwards. The problem was *not* that social science was lacking – other than in its powers of persuasion; the problem was that those voices were ignored. It should also be admitted, though, that one factor that made it easy for those who were benefiting from the deregulated free-for-all to ignore such warnings, was that there were other social scientists, especially in business schools, who were pronouncing that all was right with the world, and that the continued monetisation of the global economy – whereby the market was pushed further and further into social life, with more and more relationships and transactions becoming transferred into market mechanisms – would bring efficiency and stability.

Corporate diversity

While the global economy from the 1980s became more deregulated, market dominated, and monetised, the economy also became less diverse, with the large shareholder-owned company coming to dominate, to an increasing degree, the major economies and globally. This was particularly pronounced in the UK, which had never had the strong 'mittelstand' small- and medium-sized industrial sector as in Germany and other countries, and where the customer-owned mutual building societies were largely demutualised from the 1980s, thus further concentrating the financial services sector around the large PLC banks. Social scientists have pointed to the benefits of having a more corporately diverse economy:

> mixed institutions are always likely to be more useful for change and innovation and more resilient than those that reproduce similar characteristics over a wide institutional range. (Crouch, 2005: 59)

Following the 2007–2008 global financial crisis, there has been increased recognition that the 'biodiversity' of the economy is important, and that alongside shareholder-owned firms it is healthy to have family-owned companies, state-owned companies at local, regional and national level, and a strong co-operative, employee-owned and mutual sector.

Andy Haldane, Executive Director of Financial Stability at the Bank of England, has described well the way in which one of the factors that lay behind the 2007–2008 global financial crisis was that individual institutions had been diversifying, and that while this might be thought to reduce risk, it does not if all are diversifying in the same way, so instead the system becomes more concentrated (Haldane, 2009: 18–19). It is a classic fallacy of composition, that what is good for an individual institution acting alone does not apply when you consider all of them together. In addition to increasing risk through reduced diversity, this process also had the effect of shifting risk from the shareholder-owned banks that moved into investment banking, to the public sector, on account of the Bank of England's obligation to act as Lender of Last Resort.

The Centre for European Policy Studies produced two research studies of diversity in European banking, both of which emphasise the advantages of having diversity in banking structures and models. Their first report, *Investigating Diversity in the Banking Sector in Europe*, found that 'The most important conclusion is that the current crisis has made it even more evident than before how valuable it is to promote a pluralistic market concept in Europe and, to this end, to protect and support all types of ownership structures' (Ayadi et al., 2009: 3).

The problem is not just that the economic future is uncertain, but that it is fundamentally unpredictable. As *The Economist* notes: 'Just as an ecosystem benefits from diversity, so the world is better off with a multitude of corporate forms' (*The Economist*, 2010: 58).

Variety is the evolutionary fuel in economic development as well as in biology (as detailed, e.g., by Hodgson, 1993). Diversity is desirable across the economy, and diversity within the financial sector itself – both a variety of corporate forms and geographical dispersion, with stronger local presence – tends to support a broader variety of corporate forms in the rest of the economy which in turn enhances competition and consumer choice (Gagliardi, 2009).

The 2010 Coalition Government in the UK therefore committed itself to bringing about a greater degree of corporate diversity in the financial services sector, including through the promotion of mutuals. And the Bank of England's *Financial Stability Report* noted:

Policy action is needed to reduce the structural problems caused by banks that are too important to fail (TITF). Larger UK banks expanded much more rapidly than smaller institutions in the run-up to the crisis and have received disproportionate taxpayer support during this crisis. That reflected a misalignment of risks on TITF banks' balance sheets, due to implicit guarantees on their liabilities. (Bank of England, 2010: 11)

The UK's Ownership Commission also concluded that good ownership requires institutional plurality, stewardship governance and stakeholder engagement (Ownership Commission, 2012). Regarding public services, it has been argued that productivity and efficiency should be united in a system of performance governance, which is 'a mutual learning process of customers and service providers, in which resources and production and consumption processes of both groups are reconciled' (Gronroos and Ojasalo, 2004).

This is an area where economists could contribute more actively, making the case for greater corporate diversity, measuring the success in achieving this, and promoting it globally.

Economic complexity

To rise to the current challenges, it may be that economics itself needs a rethink. After all, Adam Smith completely changed the thinking of his day. Marx rethought economics fundamentally, presenting his analysis as a critique of political economy. And Keynes intended that his 1936 book should revolutionise the way people thought about the economy. All were successful in their endeavours, and their work in each case is the better for it. And there are no doubt lessons that can still be learned from each of them. One way in which economics today does need to rethink is to make it far more explicit that the simplified textbook models are precisely that – simplified textbook models. The problem is that while this has no doubt been done, all too often economic policies derived from those models are applied even though there is no rational basis for believing they will have the purportedly desired effects, given that they were derived on the basis of unrealistic assumptions. So what is needed in particular is an appreciation that the simplified textbook models should not be used to derive policies, without all the complexity of reality being reintroduced into the analysis before the policies are developed, finalised and applied.

And when economics is used to inform policies on broader issues such as the environment, then complexity becomes even more important and pervasive, and to fully capture these complexities, economics needs to engage in genuinely interdisciplinary and multidisciplinary research. 'Genuinely' means that economics needs to listen to and learn from other disciplines in order to work constructively with them, rather than seek to impose current economic approaches onto other areas of society, thinking that this will solve problems beyond the economic.

The term 'complexity economics' was coined by Brian Arthur of the Santa Fe Institute (Arthur, 1999), leading to the idea of economies as 'complex adaptive systems'. Summarising the literature, Beinhocker (2006) argues that this approach differs from the standard view in at least five ways:

1. *Dynamics*: economies are open, dynamic systems, far from equilibrium;
2. *Agents*: they are made up of heterogeneous agents, lacking perfect foresight, but able to learn and adapt over time;
3. *Networks*: agents interact through various networks;
4. *Emergence*: macro patterns emerge from micro behaviours and inter- actions; and
5. *Evolution*: evolutionary processes create novelty, growing order and complexity over time.

Drawing on ideas of Georgescu-Roegen (1971), Richard Nelson (2005) and others, Beinhocker (2006) argues that an ongoing process of co-evo- lution of physical technologies, social technologies (that is, institutions or ways of co-ordinating human activities) and business plans under- lies the creation of wealth in industrialised countries, notably as prop- erty rights-based market economies encourage technological and social innovations for meeting (and creating) consumer demands. He argued that this approach can inform how to enhance and spread more widely this prosperity, whilst recognising the limits imposed by human impacts on the planet's climate and ecosystems.[1]

Complexity economics draws together insights from a range of approaches that challenge conventional economic thinking, including evolutionary economics and institutional economics. The roots of complex systems thinking and its application to environmental and evolutionary issues go back to Georgescu-Roegen (1971) and Kapp (1970), and significant progress has been made in recent years via the development of ecological economics that has done much to combine the analysis of ecosystems and economic systems. One of the challenges facing the further development of research integrating economic and ecosystems is its interdisciplinary nature, drawing as it does on analysis from the natural and social sciences. Working across such disciplinary boundaries is not easy; however, meeting this challenge would appear to be essential if we are to enhance our understanding of the social costs of economic activity, most especially environmental damage. Recent research grounded in a systems approach encompassing economic and ecological systems has brought insights that can enhance and enrich thinking and policy making on these issues, and offers significant potential. Perhaps a major area of achievement has been to highlight the limitations of an over-reliance on market based instruments to the exclusion of other policy measures, such as reform of governance structures, changes in shared norms and innovation strategies.

Complexity economics seeks to draw on wider thinking on 'complex systems' by applying ideas of non-linear dynamics, heterogeneous agents, networks, emergence and evolution. Various authors have sought to apply some or all of these ideas to economic thinking. In a series of volumes from the Santa Fe Institute for Complex Systems, Brian Arthur and colleagues developed ideas of the economy as a 'complex adaptive system' (Anderson, Arrow and Pines, 1988; Arthur, Durlauf and Lane, 1997; Blume and Durlauf, 2006). This line of thinking led particularly to the application of agent-based modelling to economic problems. Potts (2000) developed an evolutionary microeconomic model, in which economic systems consist of elements interrelated by multiple connections in networks. Allen (2001) and Allen, Strathern and Baldwin (2007) developed models of firms interacting in economic markets, emphasising properties of self-organisation and adaptive learning. These strands of thinking all highlight the fact that individuals and firms, though lacking in perfect foresight, are able to learn and adapt over time, and typically interact through networks. Emergent patterns arise out of these micro level behaviours and interactions, but these are only discernible at higher systems levels. Foster (2005) identifies the dissipative, evolutionary and structural irreversibility of complex economic systems as important properties, together with the possibility that systems exist as both holistic entities and components parts. It is the connections and interactions within and between systems that lead to the emergence of complexity (Foster, 2005). This property of complex systems is highly relevant to the study of environmental sustainability that requires analysis of the connections and interactions between economic and ecological or natural systems.

The application of evolutionary thinking to economics was boosted by the seminal book by Nelson and Winter (1982), which argued that individuals and firms have 'bounded rationality' and so follow habits and routines, which evolve by a process of variation, selection and retention. Metcalfe (1997) developed this relation between evolutionary economic theory and the Schumpeterian idea of economic change occurring through periods of 'creative destruction'. Dopfer and Potts (2008) sought to develop a general theory of economic evolution, based on interactions between agents and structures at micro, meso and macro levels. Beinhocker (2006) argues that economic evolution is able to explain the explosive non-linear creation of wealth, increasing levels of variety and complexity, and spontaneous self-organisation.

Economic evolution is thus argued to be strongly path-dependent – that is, 'history matters' – and technological and institutional systems may become 'locked-in', creating barriers to the adoption of more beneficial alternatives; van den Bergh (2007) argues that greater attention should be paid to the application of evolutionary theory to environmental questions, not least in the context of system resilience, resource use, ecosystem management and growth, but also in relation to individual behaviour and environmental policy. Many of the limitations of neoclassical economics spring from the underlying model of rational choice or business decision making, with no meaningful analysis of the institutional environment in which business and policy decisions are taken. Ostrom (2006, 2007) provides significant insight into the use and evolution of governance systems to manage the commons and natural common pool resources; she found, empirically, that strategies for collective action evolve and adapt, aided by the design of appropriate institutions, thus enabling systems to escape the tragedy of the commons predicted by neoclassical economics. Similarly, Michie and Oughton (2011) show how alternative managerial, institutional and evolutionary theories provide richer insights into environmental problems and a broader spectrum of policy choices.

The need for interdisciplinarity and new economic thinking

This need for new economic thinking, and for multidisciplinary and interdisciplinary work, including an economics that is open to new thinking and to contributing to genuinely joint endeavours, has gained some support over the past few years. The World Economics Association is a new international network of economists created in 2012 to debate and discuss issues with just such an open-minded agenda.

George Soros and others have funded the institute for new economic thinking, and one of the leaders of this new initiative is Professor Sir David Hendry, cited above as being one of those who actively debated the monetarist theories of the early 1980s, and who more recently researched an impressively interdisciplinary analysis of climate change (Hendry, 2011). It is to be hoped that economics as a discipline will take this opportunity to undertake a fundamental rethink – as it has done successfully in the past, with Adam Smith, Marx, and Keynes. There is a need for an economic approach that draws on the richness of past analysis while appreciating the new complexities to be analysed, and that is able to combine with scholars from other disciplines to address the big issue of the day – now and in the future.

This chapter began with a reference to the work of Stiglitz et al. (2009) into the nature of economic performance and social progress, and a brief reference was made to their conclusions, which were that we do indeed need to take a broader view of such matters than economics has generally done to date. It is also worth drawing attention to another important point made by that large piece of work – which involved a great many people, beyond the three principal authors – namely regarding the way in which that Commission went about conducting their work, which was primarily a work of social science, and more specifically, economics. They describe it thus:

> This is a report written by economists and social scientists. The members of the Commission represent a broad range of specialisations, from national accounting to the economics of climate change. The members have conducted research on social capital, happiness, and health and mental well-being. They share the belief that it is important to build bridges between different communities – between the producers and consumers of statistical information, whatever their discipline – that have become increasingly distant in recent years. Commission members see their expertise as a complement to reports on similar topics that were written from a different perspective, for instance by scientists on climate change or by psychologists on mental health. (Stiglitz et al., 2009: 10)

Writing in 1930 on what economic life might be like a hundred years hence, Keynes was optimistic about the growing levels of production and economic possibilities that would be delivered as a result of techno-logical progress, despite the bleak economic circumstances at the time. He was perhaps over-optimistic as to what future generations would do with the newfound possibilities for leisure. But he was prescient regarding the scope and scale of what was possible and likely in terms of economics and the economy. But this did not lead him to an exaggerated importance of his profession. Quite the contrary, Keynes (1930) concluded his essay with the following plea:

> The pace at which we can reach our destination of economic bliss will be governed by four things – our power to control population, our determination to avoid wars and civil dissensions, our willing-ness to entrust to science the direction of those matters that are the proper concern of science, and the rate of accumulation as fixed by the margin between our production and our consumption; of which the last will easily look after itself, given the first three.
>
> Meanwhile there will be no harm in making mild preparations for our destiny, in encouraging, and experimenting in, the arts of life as well as the activities of purpose.
>
> But, chiefly, do not let us overestimate the importance of the economics problem, or sacrifice to its supposed necessities other matters of greater and more permanent significance. It should be a matter for specialists – like dentistry. If economists could manage to get themselves thought of as humble, competent people, on a level with dentists, that would be splendid!

Conclusion

At the time of writing (November, 2014), the UK and global economies are still only slowly and fitfully recovering from the global recession of 2009, which was the first year since the 1930s that the world's aggre-gate output, consumption and income actually fell. In that year, while the emerging economies saw their growth rates drop from an average of 9.1% in 2007 and 5.8% in 2008 to just 3.1% in 2009 – still though a positive figure, the world's developed economies saw their average growth rate of 2.7% in 2007 fall to just 0.1% in 2008 and -3.6% in 2009. This resulted in the growth rate for the world economy as a whole falling from 5.2% in 2007 to 2.4% in 2008 and –0.8% in 2009. That global recession did provoke economic debate and controversy, both amongst social scientists and between policy makers. Initially there was a co-ordinated attempt at global fiscal expansion to pull the world's economy out of recession, and this resulted in a 2.9% growth of global income in 2010, but this recovery soon gave way to austerity measures aimed at reducing the debts that had been caused by both the bank bailouts to deal with the global financial crisis and by the subsequent recession.[2]

These debates have therefore continued, both around the austerity measures (critiqued e.g., by Blyth, 2013) and around what policies should be put in place to prevent a recurrence of the 2007–2008 global financial crisis. Here the debates have focused on the need for banks to hold more capital (see Admati and Hellwig, 2013), on the need to tackle the 'too big to fail' problem (see the Vickers Report – The Independent Commission on Banking, 2011), and on the need for greater corporate diversity within the financial services sector (see Michie, 2010).

The continuing research of such issues within social science is vital to achieving the best outcomes for our economy and society. While debates within social science are inevitable and necessary, as alternative theories and hypotheses are tested against each other and against the data, and as new developments require new analysis, improved policies and performance will require continued research into the economic processes involved, and into how these can be best shaped and incentivised along a new trajectory of sustainable development.

Notes

1. See Foxon et al. (2013) where these points are analysed and discussed in detail, and on which this section draws.
2. All growth figures are of Gross Domestic Product (GDP) growth at Purchasing Power Parity and are from the Economist Intelligence Unit, cited in *The Economist* (2013: 20).

Bibliography

Admati, A. and Hellwig, M. (2013). *The Bankers' New Clothes: What's Wrong with Banking and What to Do about It*, New Jersey, USA: Princeton University Press.

Allen, P. M. (2001). Knowledge, ignorance and the evolution of complex systems. In J. Foster and J. S. Metcalfe (eds), *Frontiers of Evolutionary Economics: Competition, Self-Organization and Innovation Policy*, Cheltenham, UK and Northampton, MA, USA: Edward Elgar.

Allen, P. M., Strathern, M. and Baldwin, J. S. (2007). Complexity and the limits to learning. *Journal of Evolutionary Economics*, 17, 401–431.

Anderson, P. W., Arrow, K. and Pines, D. (1988). *The Economy as an Evolving Complex System*, Santa Fe Institute Studies in the Science of Complexity. Reading. MA: Addison-Wesley.

Arthur, W. B. (1999). Complexity and the Economy. *Science*, 284, 107–109.

Arthur, W. B., Durlauf, S. N. and Lane, D. A. (1997). *The Economy as an Evolving Complex System II*, Santa Fe Institute Studies in the Science of Complexity. Reading, MA: Addison-Wesley.

Ayadi, R., Schmidt, R., Carbo Valverde, S., Arbak, E. and Rodriguez Fernandez, F. (2009). *Investigating Diversity in the Banking Sector in Europe: The Performance and Role of Savings Banks*, Brussels: Centre for European Policy Studies.

Bank of England (2010). *Financial Stability Report*, Issue No. 27, June, London: Bank of England.

Beinhocker, E. (2006). *The Origin of Wealth: Evolution, Complexity and the Radical Remaking of Economics*, London: Random House.

Blume, L. E. and Durlauf, S. N. (2006). *The Economy as an Evolving Complex System III: Current Perspectives and Future Directions*, Santa Fe Institute Studies in the Science of Complexity. Reading, MA: Addison-Wesley.

Blyth, M. (2013). *Austerity: The History of a Dangerous Idea*. Oxford, UK: Oxford University Press.

Burchell, B. (1992). *Changes in the Labour Market and the Psychological Health of the Nation*. In J. Michie (ed.), chapter 10, pp 220–233. The Economic Legacy, 1979–1992, Massachusetts, USA: Academic Press.

Chang, H-J. (2002). *Kicking Away the Ladder: Development Strategy in Historical Perspective*, London: Anthem Press.

Crouch, C. (2005). *Capitalist Diversity and Change*, Oxford, UK: Oxford University Press.

Desai, M. (1981). *Testing Monetarism*, London: Pinter.

Dopfer, K. and Potts, J. (2008). *The General Theory of Economic Evolution*, London and New York: Routledge.

Foster, J. (2005). From simplistic to complex systems in economics. *Cambridge Journal of Economics*, 29, 873–892.

Foxon, T. J., Köhler, J., Michie, J. and Oughton, C. (2013). Towards a new complexity economics for sustainability. *Cambridge Journal of Economics*, 37(1), 187–208.

Friedman, T. L. (2005). *The World is Flat: A Brief History of the Twenty-First Century*. New York: Farrar, Straus and Giroux.

Gagliardi, F. (2009). Financial development and the growth of cooperative firms. *Small Business Economics: An Entrepreneurship Journal*, 32(4), 439–464.

Georgescu-Roegen, N. (1971). *The Entropy Law and the Economic Process*, Cambridge, MA: Harvard University Press.

Glyn, A. (2007). *Capitalism Unleashed: Finance, Globalisation and Welfare*, Oxford, UK: Oxford University Press.

Gronroos, C. and Ojasalo, K. (2004). Service productivity: towards a conceptualization of the transformation of inputs into economic results in services. *Journal of Business Research*, 57(4), 414–423.

Haldane, A. G. (2009). *Rethinking the Financial Network*, Speech delivered at the Financial Student Association. Amsterdam.

Hendry, D. (1980). Econometrics: Alchemy or Science?. *Economica*, 47(188), 387–406.

Hendry, D. (2011). Climate change: lessons for our future from the distant past. In S. Dietz, J. Michie and C. Oughton (eds), *The Political Economy of the Environment: An Interdisciplinary Approach*, London: Routledge.

Hendry, D. and Ericsson, N. R. (1983). *Assertion Without Empirical Basis: An Econometric Appraisal of Monetary Trends in the United States and the United Kingdom* by Milton Friedman and Anna Schwartz, *Bank of England Panel of Economic Consultants*, Monetary Trends in the United Kingdom, Panel Paper No. 22, October, 45–101.

Hodgson, G. (1993). *Economics and Evolution: Bringing Life Back into Economics*, Cambridge: Polity Press and Michigan: University of Michigan Press.

Kaminska, I. (2013). Beyond GDP and the rise of the non-monetised economy, FT Alphaville. Available at http://ftalphaville.ft.com/2013/04/09/1453772/beyond-gdp-and-the-rise-of-the-non-monetised-economy/, accessed 14 April 2013.

Kapp, K. W. (1970). Environmental disruption and social costs: a challenge to economics. *Kyklos* XXIII(4), 833–848; re-printed in K. W. Kapp (1974). *Environmental Policies and Development Planning in Contemporary China and Other Essays*, Paris and The Hague: Mouton, 77–88.

Keynes, J. M. (1919). *The Economic Consequences of the Peace*, New York: Harcourt Brace.

Keynes, J. M. (1925). *The Economic Consequences of Mr Churchill*, Reprinted in *Collected Writings*, London: Macmillan for the Royal Economic Society.

Keynes, J. M. (1930). Economic Possibilities for our Grandchildren. *Essays in Persuasion*, 1963. New York: W. W. Norton, 358–373.

Keynes, J. M. (1936). *The General Theory of Employment, Interest and Money*, London: Macmillan Cambridge University Press for the Royal Economic Society.

Layard, R. (2011). *Happiness: Lessons From a New Science* (2nd edn). London: Penguin.

Marx, K. (1867). *Capital: A Critique of Political Economy, Volume 1*, Moscow: Progress Publishers.

Metcalfe, J. S. (1997). *Evolutionary Economics and Creative Destruction*, London and New York: Routledge.

Michie, J. (ed.) (1992). *The Economic Legacy: 1979–1992*, London: Academic Press.

Michie, J. (2010). Promoting corporate diversity in the financial services sector. *Policy Studies*, 32(4), 309–323.

Michie, J. and Oughton, C. (2011). Managerial, institutional and evolutionary approaches to environmental economics: theoretical and policy implications. In S. Dietz, J. Michie and C. Oughton (eds), *The Political Economy of the Environment: An Interdisciplinary Approach*, Routledge Studies in Contemporary Political Economy. London: Routledge.

Natural Capital Committee (2013). *The State of Natural Capital: Towards a Framework for Measurement and Valuation*, London: Defra.

Nelson, R. (2005). *Technology, Institutions and Economic Growth*, Cambridge, MA, USA: Harvard University Press.

Nelson, R. and Winter, S. (1982). *An Evolutionary Theory of Economic Change*, Harvard University Press.

Ostrom, E. (2006). The value-added of laboratory experiments for the study of institutions and common-pool resources. *Journal of Economic Behavior and Organization*, 62, 149–163.

Ostrom, E. (2007). *The Challenge of Crafting Rules to Change Open Access Resources into Managed Resources*, Available at SSRN: http://ssrn.com/abstract=1304827, accessed 23 November 2008.

Oswald, A. (1997). Happiness and economic performance. *Economic Journal*, 107, 1815–1831.

Ownership Commission (2012). *Plurality, Stewardship & Engagement: The Report of the Ownership Commission*, London: Mutuo.

Potts, Jason (2000). *The New Evolutionary Microeconomics: Complexity, Competence and Adaptive Behaviour*, Aldershot: Edward Elgar.

Smith, A. (1776). *An Enquiry into the Nature and Causes of the Wealth of Nations*, Edwin Cannan (ed.) (1904, 5th edn). London: Methuen and Co., Ltd.

Stiglitz, J. E. (2002). *Globalization and Its Discontents*, New York and London: W. W. Norton.

Stiglitz, J. E., Sen, A. and Fitoussi, J-P. (2009). *Report by the Commission on the Measurement of Economic Performance and Social Progress*, Paris: CMEPSP.

The Economist (2010). The eclipse of the public company. London: *The Economist Newspaper Limited*, 21 August, 58.

The Economist (2013). The World in 2013, London: *The Economist*.

The Independent Commission on Banking [the Vickers Report] (2011). *Final Report and Recommendations*, London: HM Treasury.

van den Bergh, J. C. J. M. (2007). Evolutionary thinking in environmental economics. *Journal of Evolutionary Economics*, 17(5), 521–549.

7

What Can the Social Sciences Bring to an Understanding of Food Security?

Camilla Toulmin

Introduction

Food security and politics have been closely linked since Joseph gained the pharaoh's favour in early Egyptian times, by successfully interpreting his dreams.[1] Joseph forecast seven years of plenty and seven years of harvest failure. This foresight allowed the pharaoh to gather and store the surplus in good times, to take his people through hard times without famine and revolt.

Today, questions of how to ensure food security for all are often narrowly focused on increasing total food supply through, for example, raising agricultural productivity. The social sciences play a key role in questioning this emphasis, placing agricultural production in the bigger picture, and asking why, in a world of plenty, a billion people across the world still go hungry.

I take the social sciences here to include politics, economics, sociology, anthropology, human geography and psychology, these being the subjects that explore and document the functioning of our societies at macro and micro-levels. While levels of practical engagement differ, the driving purpose behind such subjects has usually been to explore systems and relationships with the aim of understanding, and seeking ways to do things better. Over the last twenty years, a conscious engagement with policy design, institutions, and decision-makers has increasingly been seen as an essential requirement by funders of research whether governmental or philanthropic. But such engagement is nothing new to the profession; eighteenth and nineteenth century writers, such as Adam Smith and Thomas Malthus, were keen to bring their insights and analysis to law-makers, debating trade, agricultural measures and food prices.

Thomas Malthus remains the writer and thinker most commonly referred to on food supply and population growth. Living from 1766 to 1834, a time of great social, economic and technological upheaval, this moral philosopher, cleric and political economist was troubled by what he saw. He is most widely known for his treatise on population growth, in which he predicted that growth in human numbers will always outstrip their capacity to produce enough food. Two hundred years later, people are still debating the same problem, having had a couple of centuries' respite from his predictions, due to falling birth rates and very rapid growth in food supply. The last of these, a consequence of opening up the great plains in North and South America, and associated transport links kept food prices low in Europe. However, the incidence of famine and malnutrition elsewhere in the world was not much affected by such technical developments, until the introduction of green revolution technologies in Asia.

Malthus also wrote cogently about both the Poor Laws (he thought they were over generous and hence encouraged too much procreation), and the Corn Laws (he thought they were essential to keep domestic grain prices high and provide a good return to Britain's farmers, though at the expense of making life tougher for those who had to buy grain). He also explored how food markets work, their tendency to swing between scarcity and gluts, with no necessary ability of the market to match demand and supply. In practice, all these issues are closely interrelated, as we see from debates about food security today. Following the 2007–2008 food price spike, global debate revolved around how to protect the poor from such price volatility, ways to combat nationalistic trade measures, and the relative merits of controlling speculation and establishing grain reserves.

A rapid review of social science and farming systems

The contribution of social science to agriculture and food security has shifted over the two hundred years from Malthus, through the colonial days when anthropologists were trained and recruited to study local people, and their societies, to the current spread of research centres and think-tanks at national and global levels. For many former colonies, in the 1960s and 1970s, there was a growing perception that social science, and economics in particular, could contribute to promoting economic development, particularly agricultural development. Following the spread of green revolution technologies, there was concern to get broader take-up of technical change, and find ways of countering resistance to change from traditional peasant farmers. Through the 1980s and 1990s, the rise of farming systems research gave a stronger role to social science, in terms of helping natural scientists to understand the constraints faced by local farmers and, hence, the options for increasing productivity and raising farm income.

A revolution in thinking took place in the late 1980s, with the growth in participatory methods, variously known as Rapid Rural Appraisal (RRA), Participatory Rural Appraisal (PRA), and Participatory Learning and Action (PLA). This mix of tools and approaches had at their heart the recognition that peasant farmers from Peru to the Philippines have a considerable body of indigenous knowledge and expertise, allied with ability and agency. Far from waiting for 'modern' science to bring answers to the problems faced, such farmers have for generations been seeking solutions, based on a mix of methods available in their context. Through the use of maps, transects, and exploration of scientific terms and concepts in local languages, research and development workers became aware of the need to understand local knowledge systems, and marry them with what insights they could bring from outside. For many schooled in traditional research and extension programmes, this meant turning things on their head, no longer privileging modern science and putting the farmer at the centre of analysis. It meant consciously listening and learning rather than telling, and 'handing over the stick'.

Reading through the early issues of RRA Notes, there is a freshness and energy evident from researchers to test out new ways of learning, and to understand how to engage a more bottom-up process of decision-making and development. It firmly rooted analysis at the level of the people who actually plan and manage their fields and landscapes day to day. It helped put 'experts' in a different place, and legitimised a re-think of whose knowledge counts.[2]

From the 1990s until the food crisis of 2007–2008, there has been less interest in agricultural and farm household economics. For many, the global food problem had been sorted, thanks to improvements in productivity. The next steps in economic development would need to move people out of farming and into urban centres and related employment, of some sort. But, since the 2007–2008 crisis, food prices and questions of food availability have come firmly back onto the global agenda, for countries rich and poor, as a result of climate change impacts, increased demand, rising fuel prices, the impact of biofuel mandates on demand for cereal and oil seed crops, and the slow rise in agricultural yields since the 1980s. In the early 2000s, a number of development NGOs argued that aid money had been diverted from the agricultural sector, and into direct budgetary support for health, education, and infrastructure, with huge cost to poor nations' ability to feed themselves. This call for more resources invested in farming was ignored until the crisis of 2007–2008 brought home to governments across the world their vulnerability to rapid price rises, and the disruptive and damaging consequences of having an agricultural sector unable to meet national needs.

Post 2007–2008, there has been an avalanche of research, analysis, and policy reports focused on whether and how the planet will be able to support a possible 9–10 billion people in 2050.[3] Each asks in its own way: what will it take to bridge the perceived gap, how should climate change impacts be addressed, and what needs to be done now to increase the likelihood of adequate food reaching the poor? Many studies seek to frame the narrative according to the insights and interests of the organisation leading the study, whether it be the Royal Society, the Soil Association, IAASTD, Syngenta, Agrimonde or FAO.

The social sciences have a particular role to play to inform and guide the flurry of ideas and arguments presented in global debates on food security, and by finding ways to work closely with those in the biological and natural sciences, so that the combined 'offer' can be better framed in a way which meets the multiple and diverse needs of farmers on the ground. Thus, social scientists need to:

- Ask difficult questions
- Assess what is actually happening, and understand why perceptions differ
- Review policy design and identify ways to intervene more effectively.

Each of these will be discussed in the sections to come, but first a look at status and whose knowledge counts.

Status and hierarchy in knowledge systems

The role and influence of the social sciences in food security debates have swung to and fro over recent decades, depending both on the discipline concerned and the broader context. The history and evolving composition of the 15 global centres of the Consultative Group for International Agricultural Research (CGIAR) gives a practical example of how this has played out over the last 30–40 years of public funding for global science on agriculture and food production. The first centres established initially by the Ford and Rockefeller Foundations focused on key commodities – wheat, maize, and rice, then tropical food systems in West Africa and Latin America. Subsequently, over the 1970s, nine additional centres were added to the original four, to bring in livestock, tropical forests, biodiversity, extension, and other food issues. Except for the International Food Policy Research Institute (IFPRI) which successfully established itself in the food policy-economics sphere, researchers in the bio-physical sciences have dominated positions of power and decision-making for decades. Economics has always sought to be first amongst the social sciences, with its status reliant on assertion of policy relevance, adherence to quantitative approaches and ability to draw on levels of theoretical abstraction which exclude others. Social scientists have been seen as secondary to the work of the centres, with smaller budgets and lower status.

There are numerous debates around what leads to hierarchy between different disciplines. Chambers argues that the highest status disciplines are concerned either with things, or they deal with people as though they were things. They tend to employ specialised and reductionist approaches, rely on precise mathematics and measurement, and become more and more knowledgeable about less and less. By contrast, low status disciplines are generalist, holistic, and deal with conditions which are irreducibly complex, diverse and unpredictable.[4]

Writing in 1982, Nyle C. Brady, Director General of the Irrigated Rice Research Institute (IRRI) says:

> Whatever influence the international agricultural research centres (IARCs) have had on food production by small farmers in developing nations has been due largely to the improved crop varieties and technologies developed by their physical and biological scientists, and, in recent years, to the inputs of agricultural economists. But we increasingly recognize that factors relating directly to the farmer, his family, and his community must be considered if the full effects of agricultural research are to be realised.[5]

Quantification is usually seen as an essential attribute in establishing an effective argument or case for action, and many people suffer from some level of quantophrenia, in which there is over-reliance on data and statistics. Eschewing statistics for their own sake, many social scientists often pursue looser qualitative, more discursive outputs and approaches which ask awkward questions, rather than offering simple solutions, which can irritate natural science colleagues. 'Anthropologists and other social scientists have often been labelled, and rightly so, as after-the-fact critics who study and report cases where change agents or technology designers have gone wrong in social, cultural or economic terms.'[6]

Status is also associated with following clearly established scientific method, which is described as replicable and falsifiable and, hence seen to be 'harder edged' and more credible than the study of society, as noted:

> One major problem the scientists on the team had was taking the anthropologists seriously, a problem that persisted throughout the project. Conventions for establishing biological conclusions stress statistical analyses of controlled experiments. Conventions for establishing anthropological conclusions involve repeated observations, cross-cultural comparison, and verification. To a biological scientist, analysis always requires quantification.[7]

Differences in language and method continue to dog relations between the natural and social sciences, and much of the gap between 'the two cultures' remains un-bridged.[8] Leon Walras, the nineteenth-century economist, adopted the methods and quantification of the natural science world to describe his analysis of market behaviour. He seized on physics, the subject which absorbed his father's intellectual energy, to establish economic models which sought to predict the actions and choices of humans in the market. This set the economics discipline on a particular pathway for more than a century. Fortunately, the scale of the mismatch in assumptions between the specification of economic models and the reality of human behaviour is now coming under serious scrutiny, with the realisation that the economy is not a closed equilibrium system, but an open disequilibrium system.[9] The assumption that human behaviour can be modelled as though each person is an independent atom with perfect information, while aiding quantitative analysis, is also increasingly seen as a terrible simplification of real life. The world in which we live is very different from this, and is better described as a complex and indeterminate landscape within which we act with imperfect information, which makes analysis messier and indeterminate.

Asking 'who sets the question?'

Much of the recent debate on global food security has been dominated by supply issues encapsulated by the frequently asked question – 'can we feed the world?'. Is there enough land, do we have the technology to grow more and at less environmental cost? How can the 'yield gap' be bridged? What changes are needed in science and technology to ensure we can feed the world? How will the impacts of climate change forecast for the next 20–30 years affect crop production in different regions? Our need for quantification means there is a strong focus on figures, with analysts arguing for an increase in total food supply of from 70 to 100%. Such statistics have become 'facts' through frequent repetition, such that it is now hard to identify the source, or question the basis on which they were derived.[10]

The social sciences have an important role to play by asking – is 'can we feed the world in 2050?' the right question? – and how might it be phrased differently? Framing of the question is a major determinant of what solutions will be considered, so it is important to consider alternative formulations and the different answers these generate. As many social and political analysts have noted, the world currently produces more than enough food to nourish our current population adequately. Hence, increasing the total volume of food available is only part of the answer to addressing hunger. Social science needs to press awkward questions which many people would prefer to ignore, such as 'why in our world are there a billion people underfed and a billion people obese?' Many of these questions relate to politics, power, and interests. It is much easier and safer to focus on science and technology as providing possible answers to questions of food security, as they would seem to provide a neutral solution, which does not threaten the status quo.

As Sen has argued, addressing hunger is fundamentally a political problem. There are strong links between economic inequality and the distribution of political power.[11] Consequently, the prevalence of hunger and famine is much more frequent in countries where government feels little accountability to the poor. Drèze and Sen remind us that:

Public action on hunger is a well-understood and well-travelled path.... The persistence of widespread hunger is one of the most appalling features of the modern world. The fact that so many people continue to die each year from famines, and that many millions more go on perishing from persistent deprivation on a regular basis, is a calamity to which the world has, somewhat incredibly, got coolly accustomed.... The eradication of famines is a fairly straightforward task... the successes achieved in different Asian and African countries in eliminating famines seem eminently repeatable in others.... Possible lines of policy are clear enough and well-illustrated by particular strategies that have been used in one form or another. Effective action is not only a matter of informed analysis but also one of determination and will, firm commitment, uncompromising resolve, and dedicated action.[12]

China's horrendous famine years of 1958–1961 illustrate only too clearly how wrong-headed policies and people's fear of speaking truth to power led to the deaths of between 30 and 45 million people.[13] There was no vehicle for public criticism, either through the political system or from the bureaucracy, both because the leadership was uncompromising, over-confident and unwilling to hear bad news, and because the consequences for the individual critic were shocking and often brutal.

Political pluralism creates a space for debate about addressing hunger, and competition for votes. The HANCI index is one tool for generating competition, and urging further action. Constructed by a coalition of development agencies, with support from the Institute of Development Studies (IDS), Sussex, this index compares 45 developing countries for their performance on 22 indicators of political commitment to reduce hunger and malnutrition, related to public spending, policies and laws. The aim is to generate knowledge of what's possible, and urge a race to the top amongst presidents and prime ministers seeking renown.[14]

The highly political nature of hunger can be illustrated by the following contrast. In late August 2013, the Indian House of Congress passed the Indian Food Security Bill, which will provide a basic subsidised food ration aimed at 800 million people, two-thirds of the population. At the same time, members of the US Congress were proposing firstly cuts to the food stamps programme, which currently helps feed more than 40 million poor people, and secondly, supporting an increase in public subsidies to a small number of large-scale commercial farmers.

Addressing hunger is not rocket-science, as Sen notes above, and much is already known about how to do it. There are an increasing number of governments that have established programmes to deliver food security for the poor, through a variety of mechanisms, such as food subsidies, employment guarantees, cash transfers, and food stamps. Presidents Lula da Silva and John Kufuor, of Brazil and Ghana respectively, received the World Food Prize in 2011, for their focus on meeting the needs of smallholder agriculture, and the set-up of social protection programmes to assure the poorest groups could access basic food security.

The World Food Prize noted:

A guiding principle for President Kufuor during the entirety of his two terms as president of the Republic of Ghana (2001–2009) was to improve food security and reduce poverty through public- and private-sector initiatives. To that end, he implemented major economic and educational policies that increased the quality and quantity of food to Ghanaians, enhanced farmers' incomes, and improved school attendance and child nutrition through a nationwide feeding program. President Lula da Silva made it clear, even before he took office as president of Brazil in 2003, that fighting hunger and poverty would be a top priority of his government. More than 10 government ministries were focused on the expansive Zero Hunger programs, which provided greater access to food, strengthened family farms and rural incomes, increased enrolment of primary school children, and empowered the poor. Zero Hunger very quickly became one of the most successful food and nutritional security policies in the world through its broad network of programs, including: the Bolsa Familia Program; the Food Purchase Program; and the School Feeding Program.

The social sciences are also posing other troubling questions which impact on food security at global and local levels, such as:

What are the governance issues around the global food supply? Why is there such a large gap between farm price and what people pay in the shops? How can a cup of coffee cost £2.30 when the beans which make it cost just a few pence? Where does power lie in the mosaic of different supply chains, and what is the distribution of revenues along the chain from field to fork? Are there ways of achieving a fairer distribution of revenue along the food chain, and if so, how?

Why is there so much waste in food systems – is it because food is too cheap? How does the pattern of waste differ between countries and contexts and what are some of the measures that could cut waste? And would cutting waste increase food availability in practice?

Does 'how' food is produced matter at least as much as 'what' food is produced? Traditional economic theory argues that it makes most sense to maximise production where marginal returns are highest and then re-distribute the product. Members of the Cairns Group of agricultural exporting nations argue for liberalisation of trade, and reductions in protection of domestic farmers. But social scientists need to ask – what would be the consequences of such dependence on food from elsewhere for the livelihoods, cultures, and politics of recipient nations? Opponents of trade liberalisation argue that food sovereignty is what matters, whereby people and nation states are able to feed themselves, rather than relying on others.

Social science also needs to ask 'who is the expert?' – how does food insecurity look to people who experience it daily? How can local and national provisions best align with people's own perceptions of need? They need to turn the topic on its head and ask – who is the real expert on food security, and can the insights and perspectives of those who face food shortages be combined with the views of global analysts?

Exploring the detail

Social science can explore the facts, describe, document, and analyse the incidence of food insecurity, by individual, household, location, and social class, to understand the pattern of food access and how it has been changing. Joint research with nutritionists has been key in understanding chronic malnutrition, as well as alerting governments and relief agencies to the onset of famine. The Indian Food Security Bill, noted earlier, has been, in part, a response to results from regular surveys which show that even today, more than 40% of children under-five are underweight, and 20% suffer from acute malnutrition. For a nation seeking middle-income status, the current Indian Prime Minister is reported to feel shame by such persistent levels of malnutrition and the country's poor ranking in nutritional league tables.

In the same way that adequate food production at global and national level is a poor predictor of food access by the poor, so too the aggregate availability of food at household level does not tell the whole story within the family. Analysis of food habits and patterns of consumption show the influence of power relations between men and women, old and young, in terms of who gets preferential access to food, especially when it is scarce. It also helps pinpoint the huge importance of getting good nutrition at particular moments, such as during pregnancy and lactation, and during the first 1000 days of life, if children are to reach their physical and mental potential.[15]

Tackling complex 'wicked' problems like obesity offers fertile ground for joint work by different social and natural science disciplines, by taking a systemic approach. The consequences of eating too much food have been on the public policy agenda for some time, given the striking growth in numbers of seriously obese people not only in North America and parts of Europe, but also increasingly in middle-income countries like India, China, Brazil and Mexico. Obesity has been linked to rising health problems, such as diabetes, heart and lung disease, arthritis and cancer, with serious consequences for the demand made on health budgets and wider social costs. Yet it seems a difficult and intractable set of behaviours to address, bringing together issues of culture, food habits, human psychology, social norms and values. Society, by failing to address good nutrition early on, pays a heavy price in the longer term. If promoting health and good nutrition took a systemic approach and were mainstreamed into government policy, this could change many aspects of interventions, such as re-thinking the built environment, encouraging walking and cycling, taxing fatty and sugary foods, and clearly labelling foods.[16]

Making sense of diverse approaches

The debate on global and local food security is often polarised around simplistic positions. The social sciences have a key role in making sense of this diversity, in opening up discussion of these different world views, and the assumptions which lie behind them. At the risk of falling into over-simplification, the argument is commonly characterised as between 'modernising technocrats' who see science as offering the solutions to scarcity,[17] and those arguing for 'food sovereignty' and models of production based on local knowledge and self-provisioning.[18] In practice, most solutions are likely to involve a combination of such extremes.

Social science needs to promote reflection on the methods, assumptions and interests associated with different world views and see how political dialogue can bridge these perspectives. Bottom-up voices are important in contesting a positivist approach to addressing food security through transformation, modernisation and social engineering. Recent reports from China describe the government's plans to shift 250 million people from rural to urban centres over the next decade.[19] This massive social engineering project seeks to integrate 70% of the country's population into urban living by 2025. A big hike in state spending is needed for the infrastructure, education, health care and pensions for ex-farmers, to achieve this shift far faster than might have happened organically. On the financial front, with most of the costs borne by local government, often by borrowing money, there are worries that a credit crisis could put such plans in jeopardy. On the human front, there are worries that such huge transformations can generate major costs for society. Humans are not just so many disconnected atoms in a giant machine, but are strongly rooted both in the biology of their evolution from ape to present day, and in the family, community and other social networks that have shaped their lives. They need to find meaning and purpose, which may not be so easy to acquire in the huge cities being built for them.

Understanding the pattern of interests, and their interplay with politics and power helps explain the conundrum of poverty and hunger in the midst of plenty. It clarifies why the obvious and right steps are often not taken. Exploring the links between party political funding, the role of lobbyists, and decisions made by government should be a more central part of social science research into food policy, from which the academic world tends to shy away, leaving the ground to policy think-tanks and NGOs. Yet a dispassionate analysis of how power works to serve a particular set of interests should be a core part of social science work, rather than being left to journalists and campaigners. Increased transparency and freedom of information should help uncover some of these connections. A recent study of US politics shows that lobbyists and politicians are closely intertwined, with an estimated 50% of former law-makers being hired as lobbyists. In 2012, there were 12,400 lobbyists spending more than $3 billion.[20] And with 4–5 lobbyists hired by oil companies for every member of congress, it is not surprising that climate change legislation is making very slow progress.

Social science also needs to show the value of expertise from different sources, in order to confer credibility on indigenous knowledge, and show that there are usually no perfect answers. Context really matters, and what works in some places, may not offer a solution elsewhere. A good example of this is seen with the orange-fleshed sweet potato in Malawi, designed to withstand drought and offer both calories and protein enriched with vitamin A. Farmers, NGOs and agricultural scientists have pooled their knowledge and resources to help spread planting material, knowledge of how best to manage and multiply the sweet potato vines, and ways to build on traditional forms of food storage and preservation.[21]

Reviewing evidence from policy interventions

Social science research and engagement have brought about major shifts in policy and practice, whether it be in assessing design of social protection measures, or analysing farm size and tenure models.

Assessing social protection measures. As noted earlier, there is plenty of evidence of the benefits from social protection measures for addressing hunger. Social science needs to assess the options in context – food distribution, income support, social safety nets, and related measures which have consequences for better nutrition, better education for girls and jobs for women.

Social protection programmes are seen as increasingly central to addressing risk and vulnerability, and enabling countries to meet the Millennium Development Goals (MDGs). In a recent study, researchers from the Overseas Development Institute (ODI) showed that significant improvements to cash transfer and other schemes are possible if attention is paid to beneficiary perceptions.[22] 'Beneficiary and community voices highlighted some key implementation challenges, including targeting flaws, payment delays, inadequate grievance channels…' In a review of five different cash transfer schemes, beneficiaries highlighted the positive aspects to be: increased control and decision-making powers, better access to basic services, improved intra-household relations, restored dignity, less dependence on others, and increased confidence. With the knowledge that a cash transfer would be forthcoming, it was easier for beneficiaries to gain credit, which also helped boost the local economy. Nevertheless, there were problems that arose from the operation of such schemes which included: tension between spouses as to how money should be spent, erosion of traditional and informal social protection systems, and a feeling from non-beneficiaries that they should be included. Overall, there was a preference for jobs over cash transfers, which suggests the importance of combining such transfer programmes with other complementary measures, such as skills training, access to credit, building greater social and economic resilience, and improving infrastructure, such as access to energy.

Debating farm size and structure. There have been long-running debates on the relative merits of small and large farm size, to which the social sciences have made major contributions.[23] They have explored their relative performance in terms of yield, employment, environmental management, incomes generated, and a range of other measures. Such analysis is often a key part of political debate around the support that should be accorded to large or small farms, and how far trade measures should recognise and protect the multifunctional character of smallholder agriculture.

Assessing different tenure models. Land titling, shape-cropping and tenancy issues have all been subject to social science enquiry, to valuable effect. Economists have led the analysis of institutional forms and contracts and the incentives they offer to land users to invest in agricultural land, shaping policy responses in many countries. Anthropologists have shown the range and dynamics of such social institutions in contexts of rapid social change.[24] From simplistic solutions promoting freehold title everywhere, research and practice have shown the wide range of responses that can provide both the incentive and the collateral for gaining credit that allows for investment in land improvements. Simple, cheap methods to secure rights of occupation and use are often a better option than formal titling, are accessible to the majority of the population, and offer sufficient security for invest-ment. This debate and the research work which lies behind it has led to a big worldwide shift in the design of land administration systems and processes for registering claims in land.[25] It has also shown the multiple routes down which such security can be assured, combining a range of methods which include recognition of customary law, and hybrid institutions for managing land administration which involve both traditional and modern procedures.

Arguing for common property resource management systems. A further good example of impacts from social science research and networking can be found in the change in policy attitudes towards common property resources (CPRs), such as grazing land, forests, and fisheries. Hardin's blunt analysis in the late 1960s of the inevitable 'tragedy of the commons' had huge influence on thinking, and supported measures to enclose and privatise common resources on grounds of efficiency and environmental sustainability.[26] But much subsequent work has demonstrated the limits to Hardin's findings, and the many circum-stances where CPR institutions have generated long-term sustainable solutions, and thus tragedy has been averted. Ostrom[27] played a land-mark role in this re-shaping of debate, recognised by her Nobel prize for Economics in 2009, alongside a broad network of others working on the study of common property, in fields like pastoral development, community management of fisheries and joint forest management. The theoretical underpinnings from Ostrom provided intellectual support and practical policy examples which led to the reversal of ideas and policy interventions in many places. For example, in the West African Sahel, many governments now recognise *conventions locales* which confer formal government legitimacy on local bodies seeking to manage a common resource, setting sanctions and excluding certain forms of use.[28] Equally, there has been a revolution in thinking around nomadic pastoral development in a range of countries,[29] such that governments – instead of forcing settlement – have passed legislation recognising the importance of herd mobility, and rights of access to grazing resources.

Conclusion

Social science often deals in the less quantitative, intangible, values-based issues that hold our society together, including food security. It requires an understanding of humans both as individuals and how they relate to wider society. Economics took a wrong turn in the late nineteenth century when Walras framed economics in the language of physics. Today, the social sciences continue to draw on the natural sciences, to help understand human behaviour, such as the insights gleaned from evolutionary biology. This has brought new perspectives to analysing individual and collective behaviour, such as the instincts for cooperation and competition, which combine at different levels and in multiple ways.

While borrowing from its models, social science also needs to contest the primacy of natural science in giving advice on the big public policy issues of the day, by demonstrating the value and relevance of its methods. This means describing the system as it is, including the interests at stake, mastering the best forms of data analysis and presentation when appropriate, and flagging issues of distribution, inequality and social justice, alongside questions of how to increase aggregate output. It needs to focus attention on the 'how' of food security alongside the 'what', by recognising that the means by which food security is achieved matters as much as the end in itself.

But it must also be recognised that many of the most interesting questions around 'can we feed the world in 2050?' actually cross disciplines, and require input from both natural and social sciences. Food and farming systems do not fit neat disciplinary boxes. Hence, seeking solutions to sustainable food security requires a combination of skills and knowledge to map the incidence and causes of food insecurity, as well as the options for addressing hunger. Ultimately, however, it's the politics which determines whether government acts on well-established evidence for measures to end hunger. Pharaoh's sleep was upset by fear of famine and the unrest this would cause. Joseph could interpret and translate Pharaoh's dreams into the setting up of food stocks to cope with forthcoming famine. The large number of reports focused on 'can we feed the world in 2050?' suggests that many powerful people are having sleepless nights. Learning lessons from the past and putting the interests and needs of the world's hungry at the forefront of reforms to food and agricultural systems could help them achieve the sleep of the just.

Notes

Camilla Toulmin is the Director of the International Institute for Environment and Development (IIED).

1. Genesis 31: vv 1–13.
2. Chambers, R., Pacey, A. and Thrupp, L. A. (eds) (1989). *Farmer First: Farmer Innovation and Agricultural Research*, London: Intermediate Technology Publications.

3. Foresight. *The Future of Food and Farming* (2011). London: UK Government. IFAD Rural Poverty Report (2011). Rome: International Fund for Agricultural Development. Agrimonde (2009). *Agricultures et alimentations du monde en 2050: scénarios et défis pour un développement durable.* Paris: Editions Quae. Royal Society (2009). *Reaping the Benefits: Science and the Sustainable Intensification of Global Agriculture.* London: Royal Society. Godfray, C. et al. (12 February 2010) Food security: the challenge of feeding 9 billion people. *Science,* 327(5967), 812–818. Conway, G. (2012). *One Billion Hungry. Can we Feed the World?* Ithaca: Cornell University Press. Emmott, S. (2013). *10 billion,* London: Penguin.
4. Chambers, R. (1997). *Whose Reality Counts? Putting the Last First,* London: Intermediate Technology Publications.
5. IRRI, (1982). *Report of an Explanatory Workshop on the Role of Anthropologists and Other Social Scientists in Interdisciplinary Teams Development Improved Food Production Technology,* Manila: IRRI.
6. Ibid.
7. Ibid.
8. Snow, C. P. (1959). *The Two Cultures,* Cambridge: The Cambridge University Press.
9. Beinhocker, E. D. (2007). *The Origin of Wealth,* London: Random House Business Books.
10. Soil Association (2010). *Telling Porkies. The Big Fat Lie About Doubling Food Production,* Bristol: Soil Association.
11. Sen, A. K. (1982). *Poverty and Famines. An Essay on Entitlement and Deprivation,* Oxford: Oxford University Press. Stiglitz, J. E. (2012). *The Price of Inequality,* New York: W. W. Norton & Company.
12. Drèze, J. and Sen, A. (1989). *Hunger and Public Action,* Oxford: Oxford University Press.
13. Yang Jisheng (2012). *Tombstone: The Untold Story of Mao's Great Famine,* London: Allen Lane.
14. Hunger and Nutrition Commitment Index (HANCI). Sussex: Institute of Development Studies.
15. Conway, G. (2012). *One billion hungry. Can we feed the world?* Cornell University Press.
16. Foresight (2010). *Tackling obesities: Future choices,* London: UK Government.
17. Ridley, M (2010). *The rational optimist,* London: Fourth Estate.
18. Pimbert, M. (2009) *Towards Food Sovereignty: Reclaiming Autonomous Food Systems,* London: IIED. http://www.iied.org/towards-food-sovereignty-re-claiming-autonomous-food-systems, accessed 12 September 2013.
19. Johnson, I., (2013). New York. *The New York Times,* available at http://www.nytimes.com/2013/06/16/world/asia/chinas-great-uprooting-moving-250-million-into-cities.html?pagewanted=all&_r=0, accessed 5 September 2014.
20. www.opensecrets.org
21. Abidin, P.E., Nyekanyeka, T., Heck, S., McLean, S., Mnjengezulu, G., Chipungu, F., Chimsale, R. Botha, B. (2013) *Less Hunger, Better Health and More Wealth: The Benefits of Knowledge Sharing in Malawi's Orange-Fleshed Sweet Potato Project,* Dublin: Hunger, Nutrition, Climate Justice. Conference: a new dialogue. Putting people at the heart of global development. Dublin April 15–16, 2013. http://www.mrfcj.org/pdf/case-studies/2013-04-16-Malawi-OFSP.pdf, 12 September 2013.

22. www.transformingcashtransfers.org
23. Feder, G. (1985). The relationship between farm size and farm productivity. *Journal of Development Economics.* 18(2): 297–313. Berry, R. A. and Cline, W. R. (1979). *Agrarian Structure and Productivity in Developing Countries.* Baltimore: John Hopkins Press. Barrett, C., M. F. Bellemare, J. Y. Hou, (2010). Reconsidering conventional explanations of the inverse productivity-size relationship. *World Development,* 38, 88–97.
24. Chauveau, J-P., Bosc, P. M. and Pescay, M. (1998). Le plan foncier rural en côte d'Ivoire. In P. Lavigne Delville (ed.), *Quelles politiques foncières en Afrique? Réconcilier pratiques, légitimité et légalité,* Paris : Karthala-Coopération Française, 553–582.
25. de Janvry, A., G. Gordillo, E. Sadoulet, J-Ph. Platteau (ed.) (2001). *Access to Land, Rural Poverty, and Public Action,* Oxford: Oxford University Press. Lavigne Delville, Ph., Toulmin, C. and Traoré, S. (eds)(2000) *Gérer le foncier rural en Afrique de l'Ouest. Dynamiques foncières et interventions publiques.* Paris: Karthala; Saint-Louis: URED, (2000). Baland, J-M. and Platteau, J-Ph., (2000). *Halting Degradation of Natural Resources: Is There a Role for Rural Communities?* Oxford: Oxford University Press. Deininger, K. (2003). *Land Policies for Growth and Poverty Reduction.* Washington: World Bank. Toulmin, C. and Quan, J. (2000). *Evolving Land Rights, Policy and Tenure in Africa,* London: DFID/IIED/NRI. Cotula, L. (2013). *The Great African Land Grab?* London: Zed Books.
26. Hardin, G. (1968). The tragedy of the commons. *Science,* 162(3859), 1243–1248.
27. Ostrom, E., (1990). *Governing the Commons,* Cambridge University Press. Ostrom, E. (2002). *The Drama of the Commons,* Washington DC: National Academy Press.
28. Djiré, M. (2004). *Les Conventions Locales au Mal,* London: IIED and Dakar: IED Afrique.
29. Behnke, Jr, R. H., Scoones, C. K. and Kerven, C. (eds) (1993). *Range Ecology at Disequilibrium: New Models of Natural Variability and Pastoral Adaption in African Savannas,* London: Overseas Development Institute.

8

Numbers and Questions: The Contribution of Social Science to Understanding the Family, Marriage and Divorce

Mavis Maclean and Ceridwen Roberts

In the Introduction to this volume we are advised that just opening a newspaper serves as a reminder of the problems, big and small, with which society is beset: 'Public issues and private troubles are as interlaced as ever', and while a response is required, 'solutions need to be based on accurate and suitable information'. This volume demonstrates what social science can offer towards identifying social issues, and offers solutions for those problems which are visible in public discourse. We would like to suggest that contemporary work in the social sciences has an additional, perhaps even more important, role to play – namely in the search not only for answers to known problems, but in going on to develop the next set of questions. As Martin Rein said in the introduction to his classic text on *Social Policy: Issues of Choice and Change*, 'what is needed in social policy is not so much good tools but good questions' for developing the forthcoming policy agenda (Rein, 1970). So we begin by looking critically at the way social science is used to identify current problems, and then go on to consider how social scientists can also contribute to formulating the next set of policy questions. We hope to demonstrate both the present and potential contribution of social science to a better understanding of society, and to the more effective development of policies for the family.

This chapter focuses on the private troubles which become public issues associated with the changing course of family life, in particular the formation and dissolution of adult personal relationships through marriage and divorce and the difficulties which arise for those concerned, particularly the children of these relationships, from these transitions.

Numbers and the need for skilled interpretation

The accessibility and quality of data on family life has increased exponentially over the last decade. It is now a simple task to 'google' online marriage, divorce and civil partnership statistics using an iPhone or iPad, and have instant access to the latest data collected and released by the Office of National Statistics (ONS), together with commentary published in journals such as ONS Population Trends. These demographic data are of high quality and easily accessible. But they are complex, and require careful interpretation. One difficulty with the application of studies of family trends and family behaviour is that everyone is an 'expert', bringing their own experience and personal values to the table. And every issue is set in an emotional context. As British society becomes more diverse in ethnicity, culture and belief, these differences add to the need for skilled interpretation of family data. We are caught in a situation where, although we are fortunate in having a wealth of information and keen media interest in reporting newly released data, this kind of dissemination may not always be grounded in the necessary expertise and knowledge needed for in-depth evaluation of the findings.

As well as academic research there is also a plethora of less reliable information, including heavily biased surveys with leading questions, biased samples and mis-analysis of data. The contribution of social science research on family issues will be limited if it becomes dominated by lay reporting of key findings. The impact can be far greater if the social scientists' voice can be heard – making clear what inference can be drawn from particular findings, taking into account sampling methods and sample sizes, taking care to differentiate between an association observed between two factors and the possibility of a causal relationship which will permit a level of prediction, and making the distinction between exploratory work which seeks to develop understanding of some matter, and explanatory work which will attempt to answer questions. To report on the success of a small qualitative exploratory study – such as, for example, an evaluation of a new pilot programme for helping perpetrators of domestic abuse – and to claim that the results could be generalisable to the population at large, is likely to be misleading and wrong.

With these caveats in mind, we turn to some of the key issues in family research which are relevant to policy debates surrounding the current state of the Family, in particular recent changing patterns of Marriage and Divorce, and their possible negative impact on children and society as a whole. We will review the social science evidence which underlies our current formulation of social problems, and ask whether other additional issues are emerging, whether any further contributions to policy development are indicated, and whether the latest available research reveals any need to modify the currently rather bleak picture.

Marriage and divorce statistics: the 'decline' of marriage

Popular understanding of the social problems linked to changing family structures can be summarised as 'the breakdown of the family', indicated by the decline and instability of marriage, and the impact of lone parenthood on the development of children and their subsequent ability to function as good citizens when they reach adulthood. On opening a newspaper during the party conference season of autumn 2012, we found James Chapman's headline in the Daily Mail, 8 October 2012, which runs 'The Married will be a minority in a generation...half of all children now can expect to see their parents separate'. The article draws on a report published on 7 October 2012 by the think tank the Centre for Social Justice, (CSJ) set up by Ian Duncan Smith, entitled 'Transforming Child Care, Changing Lives: Making sure work pays' which warns of an increasing social divide in family structure into traditional stable families and a large and growing less stable group who lack resident fathers.

This view was endorsed at the Conservative party conference with warnings from a former Minister for Children, Tim Loughton MP, that without concerted action Britain was in 'peril, socially and economically...Family breakdown costs society £44 billion a year'. The remedy proposed was a ten point plan which would reward marriage through the tax and benefit system, introduce a full presumption of shared parenting after separation in the family courts to maintain attachments between separated fathers and children, deregulate prohibitively expensive childcare, make practical suggestions to protect children from online pornography, and enable recently retired people to offer guidance to fatherless boys. The CSJ study found that only half of new young parents on low incomes were married, and that this proportion rises to 80% for those earning less than £21,000 a year. The report ended by concluding that by 2047 less than 50% of families would be headed by married couples.

This CSJ report continues the analysis first presented by the think tank in a series of reports entitled 'Breakthrough Britain', summarised in 2006 (Centre for Social Justice, 2006). The waning of marriage and the decrease in the traditional stable two-parent form of child rearing is firmly placed at the heart of British troubles ranging from poverty, unemployment and welfare dependency, to crime and social disorder. This kind of analysis is not new. It has characterised popular and political discourse for over thirty years and in the 1990s was chiefly associated with publications from the Institute of Economic Affairs (e.g., Dennis and Erdos, 1992).

Marriage in the latest CSJ report is described as creating citizenship and nationality, legitimacy, regulating inheritance and welfare benefits, and much more. This approach has been constructed rather differently in academic research from the American demographer Andrew Cherlin who described marriage as a 'Super relationship' (Cherlin, 2004) whose 'symbolic significance has remained high and may even have increased. It has become a marker of prestige and personal achievement'. But, as he and British social scientists such as Kath Kiernan have shown, it is increasingly associated with the more highly educated and occupationally and economically advantaged. Why is this?

Attitudinal data from the British Social Attitudes Survey show over time how people's general attitudes to marriage and cohabitation have become less traditional and more accepting of people living together outside marriage, yet national data also show that for many cohabiting couples marriage is still an aspiration.[1] The British Household Panel Survey in 2007 reported that 75% of couples under 35 currently cohabiting wanted to marry.[2]

National statistics clearly show that the frequency of marriage is declining, and the media and some politicians and pundits make a causal link between this decline and a multiplicity of social ills. But we argue that the value of the social science contribution lies not only in providing the numbers, but also in careful interpretation. Social scientists can encourage a closer look at newly published numbers, but can then place the new data in the context of other forms of trend data as well as more qualitative social research into family matters, in order to develop and extend our thinking about the phenomenon, and in so doing begin to raise new policy questions.

For example, marriage has become less prevalent insofar as a smaller proportion of the British population is currently married than in the last decade of the twentieth century, and the composition of the married population is changing.[3] But is this a decline in marriage or a rejection of marriage per se? Does this change threaten social stability, or does it provide a role model for marriage, which, though less prevalent, may be more respected than when marriage was more common? Social scientists scrutinise the demographic data, but can also explore the meanings people attach to their behaviour to explain what is happening to family life and the values given to marriage.

Historians too have contributed to this debate, raising questions about the traditional view of the frequency of formal marriage in the past. Rebecca Probert, a legal historian, questions the assumption that cohabitation was common in the seventeenth to nineteenth centuries, and argues that formal marriage was the norm at that time (Probert, 2012). It is clear, however, that the incidence of formal marriage declined from the 1960s onwards, divorce rates increased, and cohabitation began to emerge as a statistically significant family form in the 1980s. The number of marriages fell from 480,285 in the UK in 1972, to just over 270,000 in 2007, and after a further dip rose slightly to 241,000 in 2010, which gives a current marriage rate of 8.7% people marrying per thousand unmarried population aged 16 and over.[4] The highest numbers of marriages appeared among men and women in their late twenties, and two-thirds of marriages were civil ceremonies. And whereas in 1940 over 90% of marriages included both partners marrying for the first time, by 1996 this had fallen to 58% and by 2006 to 39%.

The numbers derived from national data and analyses are clear. Social scientists are telling us that there were half as many marriages in 2010 as there were 40 years earlier. The immediate concerned response is understandable. To those who see marriage as the cornerstone of social and economic stability, the place where the next generation is nurtured and socialised, this finding is frightening. But social scientists have a great deal more to offer in response to this information. They may suggest that some of this reduced incidence of marriage is because young people are marrying later for a whole variety of reasons. Similarly as people are living longer, more are experiencing longer periods of widowhood. So in assessing the 'health' of marriage as an institution it is important to not only look at the current proportion of married people but also at who will and who has 'passed through' marriage including those who have divorced. What this then suggests is that marriage is still a majority experience, though for some not lifelong.

Social scientists also ask, what are people doing if they are not marrying? Are they forming independent households alone or living with friends? Are they staying at home with their parents for longer periods? Or, as we suggest in the following section, are they cohabiting?

Rise of cohabitation

The demographic data indicate a parallel increase in cohabitation. Eva Beaujouan and Maire ni Brolchain (Beaujouan and ni Brolchain, 2011) have shown through retrospective histories taken from the General Household Survey 1979 to 2007 that in the early 1960s less than 1% of adults under 50 years cohabited, while the figure for 2007 is 17% (one in six). Social attitudes have changed, and they quote the widespread view expressed in a recent Social Attitudes Survey that marriage is 'only a piece of paper'. The early studies of these cohabitants are reported as finding three distinct groups, a small group consisting of those choosing to reject the social and legal obligations of marriage on principle, a group of those who live together early in their relationship possibly as a precursor to marriage which might follow the birth of a child, and a group living together after a divorce and reluctant to become involved with the kinds of legal obligations and regulation they had experienced in the divorce process. As a whole, the population of cohabitants was younger and less settled in work and housing than those who married. As social attitudes to living together and even having children outside marriage relaxed, the cohabiting group has come to include some of those young couples who in earlier years had married as a result of social pressure, especially if a baby was expected. But cohabitants remain a difficult group to study, as definitions of cohabitation vary, and there is no official and visible notification of change of civil status for cohabitants as there is for those who marry and are registered as doing so. As pressure to reform the law to give more of the financial protections available to women and children on divorce to separating cohabitants has developed, it has always foundered on the difficulty of evidencing cohabitation in a clear way (see Law Commission, 2007).

Since the 1990s, cohabitants have become an increasingly heterogeneous group. For some it is a temporary phase – they either marry their partner or separate; the average length of cohabitation has increased and children are more likely to be born into the relationship. For some, cohabitation has become an alternative way of family life rather than a precursor to marriage. But the socio-demographic characteristics of people making different choices vary significantly. Long-term cohabitation, especially with children of the relationship, is disproportionately associated with lower education, occupational and income levels. Cohabitation with children is less stable as a family form than marriage, as recent analysis of the Millennium Cohort has shown (Holmes and Kiernan, 2010).

Increase in relationship breakdown

The fall in the rate of marriage has been accompanied by a rise in divorce. In 1929 there were only 3,400 divorces in England and Wales. The number doubled during the 1960s from 24,000 to 56,000 in what became known as the divorce epidemic, and by the end of the 70s in 1979 there were over 127,000 divorces.[5] The increase in numbers was exacerbated by the redivorce of couples one or both of whom already had a previous marriage which had ended in divorce. But as the divorce rate fell, following the decrease in marriage rates, in 2007 it reached its lowest level since 1981, at 11.9 per thousand of the married population. By this stage the press was reporting 'divorce rates plunging as UK men discover commitment' (*Daily Mail* 15 September 2012), though no evidence was given to support the claim that this was caused by an increase in commitment.

If we turn back from the media coverage to the demographic data, it is clear that the risk factors for divorce included low age at marriage and an unexpected change in personal economic circumstances. The risk of divorce also increased with the number of children for whom the couple were responsible (Chan and Halpin, 2008). Given these risk factors for the breakdown of marriage, and given the lower age and increased economic vulnerability of cohabiting couples, the research findings that the latter are twice as likely as married couples to separate is unsurprising (see Boheim and Ermisch, 2001). However, as cohabitation has become more commonplace across all sectors of society, pre-marriage cohabitation is no longer associated with a higher propensity to divorce for those who go on to marry.

Interpreting marriage and divorce statistics

The availability of high quality demographic data on marriage and divorce has contributed to the rise of considerable public concern about the collapse of the family, and even of society, which we noted reaching a peak in the latest CSJ Report 'Transforming Childcare' (CSJ, 2012) which aims to stimulate the development of family policy, and is supportive of traditional family values. Such concerns are not new. Pat Thane in her report for the British Academy entitled 'Happy families?' published in 2010 (Thane, 2010) quotes Benjamin Disraeli in 1845 saying that 'There are great bodies of the working classes in this country nearer the condition of brutes than they have been at any time since the Conquest'. The 'Breakthrough Britain' summary report published in

2006 by the CSJ takes its place in a long line of such anxious comments. All governments since the 1990s have focused on family breakdown and have instigated policies aimed to deal with this. The Labour government emphasised parenting and family support as well as early intervention to help vulnerable families and measures to help with relationships. This has been continued by the present Coalition Government, including developing relationship support programmes and information hubs to support parents.[6] But the contribution of social science to this debate can be taken further, and perhaps a more nuanced interpretation offered.

We may begin by setting the British figures in a wider context, and by asking a broader range of questions. The British trend towards less marriage and more divorce can be seen in other European countries at this time. Should we question what the incidence of formal registration of marriage indicates? Or even whether marriage is an automatic good? If marriage is more common among the more economically stable, or among particular ethnic or faith communities, might it be socially divisive to have one form of family organisation for some groups and not for others? If cohabitations are becoming more lasting and stable, is the difference between marriage and cohabitation becoming less important for the policy maker? And what would be a 'good' divorce rate? Zero? Not necessarily. The lack of divorce provision in Eire was a serious problem until recently resolved. In Poland in the early 1990s, there was concern about the divorce rate being too low in a Catholic Country with an acute housing shortage and problems with alcohol consumption and resulting domestic abuse. In that context, there was a concern that too many women were unable to leave an abusive situation due in part to the position of the Church in rejecting divorce, and also the lack of any accommodation to move to (see chapter 1 of Weitzman and Maclean, 1992).

In the United States, William J. Goode described divorce as a consumer good, indeed a luxury item (see chapter 2 of Weitzman and Maclean, 1992). He described the rising divorce rate in the US, where, as in the UK, the petitioners are largely women, as reflecting the increased choices available to women who were taking a larger role in the labour market after the Second World War, often working full time, and able to control their own fertility through the contraceptive pill. Though they suffered an economic penalty on divorce, they were able to survive. Many of the women interviewed by Weitzman (chapter 14 of Weitzman and Maclean, 1992) described themselves as being economically worse off but happier after divorce as they controlled their own budget, even if it was diminished.

The rise in cohabitation in the UK has been observed to be associated with instability, early breakdown, and poorer outcomes for the children of these relationships. But many of these characteristics are associated with the relative youth and lack of resources of the cohabiting couples compared with those of the married couples here. These youthful relationships are associated with poorer outcomes, not because of their civil status, or the lack of a wedding ring, but because these young people are at a less stable period in their lives (Ferri and Smith, 2003). Those who choose to marry in England and Wales at the present time are older, richer, and more likely to be employed than their cohabiting peers. If age and income are controlled, the positive effect of marriage on outcomes for children is diminished (Goodman and Greaves, 2010). The married are now a different group from the cohabitants by selection. Encouraging the spread of marriage to other groups in society would reduce this selection effect and the impact of parental marriage on children could diminish or even disappear. It was interesting to note that in America, where marriage is more generally considered of high importance, marriage for the young often has a shorter duration than cohabitation in England and Wales (Cherlin, 2011).

Relationship breakdown and the impact on children

The demographers tell us clearly that the numbers of children being born outside marriage and the numbers of children experiencing parental separation are increasing. It would be easy to make a leap in interpretation from these figures to an assumption that family life is no longer offering children the stability they need to develop into responsible and productive citizens. But again social scientists from the fields of family sociology and developmental psychology can add a more detailed understanding of these social changes to the picture provided by the demographers. For example, we should not equate birth outside marriage with birth outside a stable parental or family relationship. Many cohabiting parents can and do provide long-term stability and care. The higher rate of breakdown of cohabiting relationships is clearly linked to youth and poverty, and if these factors are controlled cohabitation per se, loses its negative impact on outcomes for children (Rutter, 2010). We no longer automatically equate divorce with poorer outcomes for children.

Where children are exposed to parental conflict or domestic abuse within a marriage there is a case to be made for the beneficial impact of divorce, if parents can cooperate in supporting the ongoing relationships with both parents whenever feasible and safe. The government at the time of writing (October 2014) has recently legislated to encourage cooperative parenting after parental separation or divorce, by amending the Children Act (1989) to indicate clearly that involvement with both parents is likely to benefit the child.[7] It may be that the negative impact of divorce in the days when it frequently led to poverty for the divorced mother and her children, as well as little contact with the father, may in future be mitigated by better arrangements for ensuring financial obligations are met, and by supporting the child's relationships with both parents when they are able to cooperate, even though they are not and may never have been a parental couple, and even taking into account the economic burden of maintaining two households. Judy Dunn's work indicates the ability of most children to cope with transition and change in their family structure, though warns against the damaging effects of multiple transitions (Dunn, 2005). The increase in break up of first families has been accompanied by the increase in reconstituted families, where a step-parent joins the lone parent. The social science literature shows how repartnering offers a way out of poverty for the lone mother, but also highlights the difficulty of blending families where both parties bring a child to the relationship and this is increased many times over if they go on to have a child of the new relationship.

More women are now giving birth outside marriage, but the proportion who do not register the child in the names of both parents is low (7% in 2010) and decreasing. Lone parenthood generally lasts no more than five years, and fathers are far more visible both socially and legally, though the millennium study shows worryingly that a significant percentage of new fathers dropped out of sight within a year (Kiernan and Smith, 2003). Joint birth registration carries with it the legal status of parental responsibility under the Children Act England and Wales 1989, and the financial responsibility of all parents, regardless of their civil status, to all their children is firmly stated in the Child Support Act 1991 (though enforcement of this obligation remains a taxing issue here as in other jurisdictions). So while social science contributes data, and the media tends to respond with rather hasty or simplistic assumptions about the implications, we must rely on social scientists to make their more complex and nuanced observations in response to the empirical data heard better. Their message is that we are seeing changes to family structures and child rearing practices, rather than the dangerous death of family values, and a move towards flexibility and diversity rather than rigidity and conformity.

The reemergence of fatherhood

IPPR Social Policy Paper no 1, 'The Family Way', published in 1990 and comparing the way the family has been addressed on the right and left of the political spectrum – drawing extensively on social research – makes little mention of fathers. There are sections on changing patterns of family life, marriage and divorce, lone parents, remarriage and the impact on children. But no heading includes the word 'father'. It is hard to imagine any similar document being published today without a substantial discussion of fatherhood. The reemergence of fathers as a vocal group demanding the right to a continuing meaningful relationship with their children in the event of parental separation, supported by the coalition government's emphatic commitment to 'cooperative parenting' in the event of relationship breakdown is now highly visible. Lack of a paternal role model has been seriously considered as a factor in the social unrest in London in 2010 and as a factor contributing to the culture of reliance on benefits rather than employment, and to lack of attainment in education.

What is social research currently contributing to the analysis of these social issues? The impact is clearly visible in the recent debate about cooperative or shared parenting. We have so far been looking at how social science can extend and develop discussion about current social problems in a broad sense. Here we can see the contribution of social science to a high-profile public debate in a more specific and focused way. Groups of separated fathers began to challenge the traditional arrangements made on divorce whereby typically children would see their father once during the week, and spend a day at the weekend or alternate weekends staying with him plus additional time in the school holidays. These fathers' groups began to argue for shared parenting or equal parenting time, and argued that in the cases where disputes between the parents could not be settled except by going to court, that the family courts were secretive in character and biased in their decision making towards the mothers (who were also more likely to have legal advisers under the legal aid scheme as they were more likely to be financially eligible). Government research into the matter, however, showed that the fathers applying for contact or residence orders were usually successful (Hunt and Macleod, 2008).

Some of these groups, for example Fathers4Justice, adopted extreme tactics – including scaling the walls of Buckingham Palace and the House of Commons – to publicise their cause. Others worked more quietly to increase public awareness of what they saw as the exclusion of fathers from their children's lives after parental separation. This kind of activity also developed in other jurisdictions – Australia, where fathers known as the Black Shirts picketed the homes of women thought to be refusing contact, and France, which saw the SOS or Secour au Separes. The issue gathered notoriety, and in England the fathers initially received sympathetic press coverage. Politicians took an interest, and began to offer support. Finally the Coalition Government in 2012 set out its proposals to legislate for cooperative parenting by amending the Children Act 1989 to include a presumption that unless there was evidence that harm to the child might result, that any court must assume that continuing involvement of both parents with the child would further the welfare of the child.

Similar legislation had been passed in Australia in 2006, but has not passed without criticism. Extensive evaluation studies by socio-legal researchers in Australia found that the drafting of their provisions had led to confusion, and in particular to fathers believing that they had a right to half of the child's time and that this had led to difficult court cases.[8] There was also concern that there were not adequate safeguards for the child in cases of domestic abuse. The Review of Family Justice chaired for the government by David Norgrove in England and Wales recommended against further legislation.[9] The report draws on the work of social scientists in a number of jurisdictions, saying that there was no need to risk the negative impact of legislation when the courts, under the Children Act 1989, already had the welfare of the child as their paramount consideration in making any decision about the child (the strongest possible legal imperative) and that it was already widely accepted that a continuing relationship with both parents was in the best interest of the child. Any legislative change would risk putting the interests of the adults involved above those of the children, which are carefully and clearly protected in the present legislation. We have here an example of a political imperative to act, countered by a clear research-based argument for retaining the status quo. Academic arguments were carefully summarised, and communicated in various briefing documents (see note 8). Oxford Family Policy Briefing Paper 7 reported in summary form research which had examined in what circumstances shared care worked well for children, and in what circumstances it worked less well. It was clear that where parents could work together, flexibly, putting the child's needs first, that this could be an excellent arrangement. But where the parents were highly conflicted, where there was domestic abuse, or where they could not work together, the children did not do well. Neither social science, including the work of socio-legal scholars, developmental psychologists and family sociologists, nor the difficult experience of Australia could dissuade government from choosing to legislate for more father involvement with children after divorce. The Children and Families Bill had its first reading in the House of Commons on 5 February 2013. But research had filled out the picture and has lead to a far more nuanced and helpful way forward. Instead of a law which divides the child's time equally between mother and father, we have a proposal for continuing involvement with both, where this would not only not give rise to risk of harm to the child but would positively further its welfare. No particular form of contact is recommended, whether direct contact, supervised or indirect contact. The judge continues, under the Children Act 1989,[10] to put the welfare of the individual child in these circumstances at this moment in time in first place. Perhaps there was no need for this legislative proposal. But, given that politically it had momentum, this provides a clear example of the impact which a full and complex message from a large number of studies from different disciplines can have in achieving an important and helpful modification of a rather blunt instrument.

Concluding observations

This chapter in considering how social science can contribute to problem solving in the field of Marriage and the Family began with the hard evidence of demographers about who marries and who divorces, and argued for regarding this evidence as offering more than just numbers. We cautioned against hasty interpretation which does not take into account the 'softer' evidence of qualitative detailed work on family behaviour from family sociologists and psychologists necessary to develop a more multi-dimensional account of what lies behind current trends to fewer marriages and more cohabitation. This multi-dimensional account cannot be complete without considering the work of labour market economists and its implications for approaching the issues of work life balance presented elsewhere in this volume, in Chapter 1 nor without the work of those who study child development and parenting, which also forms an integral part of the contribution of social science to problems, perceived or hypothesised, associated with modern family life. We close with a plea for continuing access to high quality demographic data to be accompanied by careful interpretation and contextualising in order that social science may make the fullest possible contribution – not only by responding to known problems, but also in seeking out the issues to come.

Notes

1. See British Social Attitudes Survey 2006.
2. 'Marriage still the ideal for many couples currently living together', in Honourable Intentions? Attitudes and Intentions among Currently Cohabiting Couples in Britain by Dr Ernestina Coast. This paper was presented at the British Household Panel Survey 2007 conference at the Institute for Social and Economic Research| (ISER), Essex, accessed at http://www.lse.ac.uk/collections/pressAndInformationOffice/news/AmdEvents/archives/2007/
3. ONS Marriages in England and Wales, see footnote 8.
4. ONS Marriages in England and Wales 2010, Statistical Bulletin released February 2012.
5. See ONS Social Trends 2009, 39: 20; 2004, 35: 4; and 2008, 38: 7.
6. Supporting separated families: securing children's futures, Cm 8399, DWP, July 2012 followed by the web app SOS, Sorting out Separation, located on other relevant websites from November 2012.
7. Cooperative parenting following family separation: proposed legislation on the involvement of parents in a child's life, Department of Education, London, November 2012.
8. For a summary of the research findings and policy implications see Family Policy Briefing 7, University of Oxford Department of Social Policy and Intervention, 'Caring for children after parental separation: would legislation for shared parenting time help children?', May 2011, ISBN 978–0–5623–2–5.
9. Family Justice Review, Final Report, November 2011, Ministry of Justice and Department of Education.
10. See the Children and Families Act 2014 which amends the Children Act 1989 to support continuing parental involvement but subject to the welfare paramountcy principle

Bibliography

Beaujouan, E. and ni Brolchain, M. (2011). Cohabitation and marriage in Britain since the 70s. *Population Trends*, 145, ONS.

Boheim, R. and Ermisch, J. (2001). Partnership dissolution in the UK: the role of economic circumstances. *Oxford Bulletin of Economics and Statistics*, 632,197–208.

British Social Attitudes (2009). A.Park, C Bryson, E Clery, J Curtice and M Phillips (eds) NatCen, see http://www.bsa-30.natcen.ac.uk/media/37580/bsa30_full_report_final.pdf

Centre for Social Justice (2012). *Transforming Child Care: Making Sure Work Pays,* CSJ Report.

Centre for Social Justice (2006). *Breakthrough Britain* (Summary). CSJ Report.

Chan, T. W. and Halpin, B. (2008). *The Instability of Divorce Risk Factors in the UK,* Oxford: University of Oxford.

Cherlin, A. (2011). Studying the family, paper to Nuffield Foundation London, 1 July 2011.

Cherlin, A. (2004). The deinstitutionalisation of American marriage. *Journal of Marriage and the Family*, 66, 848–861.

Clarke, L. and Roberts, C. (2011). Family structure and family policy and practice. In M. Wadsworth and J. Bynner (eds), *A Companion to Life Course Studies*. Abingdon: Routledge.

Dennis, N. and Erdos, G. (1992). *Families Without Fatherhood*. London: IEA.

Dunn, J. (2005). Demographic change and the changing role of the family, paper to the British Academy Seminar 'Addressing the Socio economic Agenda', London, November.

Ferri, E. and Smith, K. (2003). Partnerships and parenthood. In E. Ferri, J. Bymmer and M. Wadsworth (eds), *Changing Britain, Changing Lives*, London: Institute of Education, 105–132.

Goodman, A. and Greaves, E. (2010). Cohabitation, marriage and child outcomes. *Institute of Fiscal Studies Commentary C114*, London: IFS.

Holmes, J. and Kiernan, K. (2010). Fragile families in the UK: evidence from the Millennium Cohort Study. York Working Paper, available at http://www.york.ac.uk/media/spsw/documents/research-and-publications/HolmesKiernan2010, accessed, 10 November 2014.

Hunt, J. and Macleod, A. (2008). *Outcomes of Applications to Courts for Contact Orders after Parental Separation or Divorce*. London: Ministry of Justice.

Kiernan, K. and Smith, K. (2003). Unmarried parenthood: new insights from the millennium cohort study. *Population Trends*, 114, 26–33.

Law Commission (2007). Cohabitation: the financial consequences of relationship breakdown, *Law Commission Paper* 307, London.

Social Trends (2004) (eds) Summerfield, C and Babb, P , London, TSO, also www.ons.gov.uk/ons/rel/social-trends-rd/social-trends/no--34--2004.

Social Trends (2008) (eds) Self, A and Zealey, L Office for National Statistics, Palgrave Macmillian, Basingstoke, also www.ons.gov.uk/ons/rel/social-trends-rd/social-trends/no--38--2008.

Social Trends 39 (2009) (eds) M. Hughes, J. Church and L. Zealey; Palgrave Macmillan or social_trends_39_tcm77–137023, www.ons.gov.uk/.../social-trends-39/social-trends-full-report.pdf.

Probert, R. (2012). *The Changing Legal Regulation of Cohabitation*, Cambridge: Cambridge University Press.

Rein, M. (1970). *Social Policy: Issues of Choice and Change*, New York: Random House.

Rutter, M. (ed.) (2010). Family structure, chapter 2, *Family*, p. 90. London, British Academy.

Thane, P. (2010). *Happy Families*, London, British Academy.

Weitzman, L. and Maclean, M. (eds) (1992). *Economic Consequences of Divorce: The International Perspective*, Oxford: The Clarendon Press.

9

Crime, Policing and Compliance with the Law

Mike Hough

Introduction

This chapter focuses on the contribution that social scientific research has made to our understanding of crime and its control. As a British criminologist, my focus is on Anglophone criminology, which means – largely but not entirely – Anglo-American work. I have attempted not to be parochial in drawing solely on the British experience. As a discipline (or perhaps a sub-discipline, or a multi-disciplinary fusion of the sociology of deviance, the psychology of offending and criminal law) criminology is a fairly recent invention, which can be dated to the 1950s in the UK and the US. I shall argue that its impact on academic understanding of the issues has been substantial. Until the immediate post-war period, the police, prosecutors and judiciary in industrialised countries were hidden effectively from research scrutiny. I shall summarise developments since then in three areas of criminological research:

- crime trends and social indicators of crime
- police work and the impact of the police on crime
- the role of normative compliance in explaining conformity with the law.

It is hard for someone starting a career in criminology today to appreciate the full extent to which academic knowledge has developed over the last five or six decades. For all this achievement, however, the impact of academic work on criminal justice *policy* remains quite marginal. In part this is because – in the UK context at least – the centre of gravity of academic criminology for much of this period has been characterised by critical commentary.[1] But academic reticence to engage with policy is only part of the story: at the same time that research has made great progress, 'law and order' has become an increasingly politicised issue, notably in the UK and the US, but also in some mainland European countries. The status of academic criminologists in government in the corridors of power has been in decline, and the voice of the 'academic expert' is only one of many, and quite a small voice, in public and political debate about crime. The final part of this chapter discusses the pressures on politicians to offer populist solutions to the problems of crime and disorder, and to misread or ignore what social scientific research has to tell them.

Inevitably I have been selective in focusing on three themes within criminological research, and in doing so have drawn on my own professional interests and experience. I am *not* claiming particular significance for research that addresses these themes. Rather, the research described here is intended to be illustrative both of the contribution made by criminological research – and of the factors that limit its reach into public and political debate.

Crime trends and the measurement of crime

Criminology is almost by definition an applied discipline, as the object of study – crime – is constructed by social institutions and it would be hard to engage in criminology without discovering *some* impulse either to critique or try to improve the functions of these institutions. Until the 1960s, however, there were large gaps in knowledge about crime. The most significant of these was the lack of any firm knowledge whatsoever about the extent of crime, and the proportion of *crimes committed* that get *reported* by victims to the police, and the proportion of these reported crimes that actually find their way into police statistics. Statisticians from Quetelet[2] onwards had been aware of this, but lacked any viable technology for estimating the 'dark figure' of unrecorded crime.

Edward Troup's preface to *Criminal Statistics of England and Wales, 1894*, reads as a strikingly contemporary account of the limitations of statistics of crimes recorded by the police (Home Office, 1896, quoted in Morris, 2001), anticipating more recent commentators, notably Kitsuse and Cicourel (1963):

> Not only do the figures fall short of the real number of *crimes committed* by the enormous number of unreported or unknown cases; but there seems much reason to think that, though the instructions as to the mode of collecting them have been made as definite as possible,

there is still a tendency on the part of some police forces to adopt a very high standard of what constitutes a *crime committed* or a *crime reported to the police*, and by this means further to reduce the number of cases entered into this column ... no doubt it is natural ... that they should seek to minimise the amount of unpunished crime existing in their district, but such a tendency detracts so much from the value of the returns of crime that it almost raises the question whether it is worth retaining the returns at all ... it should be clearly stated that they represent only the crimes known to the police, and do not even approach the real total of crime.

The 1965 US President's Crime Commission marked the start of a step-change in our understanding of crime levels and trends. Commissioned by President Lyndon B. Johnson, this was a response to the growth in public concern about crime in America. It was an enormously well-funded enterprise, and from the outset relied heavily on academic expertise to assemble reliable evidence about crime. The evidence-gathering process was wide-ranging, but included a national 'victim survey' of 10,000 adults supplemented by three city-level surveys in Washington, Chicago and Boston. These surveys asked representative samples of the population about their experience of a range of different crimes, and by aggregating up to the (national or city-level) population could derive estimates of the extent of crime independently of statistics collated by the police. They showed that non-reporting of crime was extensive, and that the 'dark figure' of unrecorded crime was for most offence groups very much larger than the police count.

Partly on the basis of these snapshot surveys the Commission decided that it was hard to draw firm conclusions about trends in crime from crime statistics collated by the police as it was 'likely that each year police agencies are to some degree dipping deeper into the vast reservoir of unreported crime' (President's Commission, 1967: 30). The Commission recommended a programme of research into surveys of victimisation which led to the establishment in 1973 of the National Crime Survey (NCS, redesigned and relabelled in 1993 as the National Crime Victimisation Survey – or NCVS). The NCS/NCVS was the first large-scale national crime survey that yielded reliable estimates of crimes committed against people and their private property *independent of police statistics*.

The concept of crime surveys took hold quite quickly in other countries. The Dutch Ministry of Justice (Research and Documentation Centre) launched a national survey in the early 1970s. In the UK, the

Home Office commissioned the Cambridge Institute of Criminology to carry out a crime survey in London in 1973 (Sparks et al., 1977), and launched a national survey in 1982 (Hough and Mayhew, 1983, Hough and Maxfield, 2007). The British Crime Survey, initially covering England, Wales and Scotland, relied initially on the advice and support of those academics who had been involved in the US programme of survey work (notably Al Biderman, Richard Block, Al Reiss and Wes Skogan) but also drew on the experience of researchers in the Dutch Ministry of Justice, Lesley Wilkins, David Farrington and others. The survey gradually became institutionalised, growing in scale and frequency; it was relabelled the Crime Survey for England and Wales (CSEW) in 2012, reflecting the fact that separate though similar surveys are now mounted in Scotland and Northern Ireland. Crime surveys have now been carried out in most European countries, as well as in Australia and New Zealand, and in a range of developing countries. Many of these have used the standardised questionnaire and administration methods of the International Crime Victimisation Survey, designed specifically to enable cross-country comparative research (see e.g., van Dijk et al., 2007).

These initiatives were, for the most part and especially the US NCVS, in the tradition of social indicators research, rather than exercises in theory-building and theory-testing. Some of the surveys – notably the BCS/CSEW and the Dutch programme of crime surveys – were however conceived of as survey *research*: information was collected not just on respondents' crime experience, but also on their social and economic status, the types of neighbourhood they lived in, the patterns of their everyday 'routine activities', their attitudes to crime and punishment and their concerns about crime, and their ownership and use of crime prevention technology. Over the years, an increasingly fine-grained picture has emerged of variations in vulnerability to crime and the reasons for this.

However the most valuable contribution made by crime surveys probably remains their ability to disentangle changes in levels of reporting and recording of crime from the underlying trend. Figure 9.1 uses the CSEW to exemplify this. The top line in Figure 9.1 shows trends in crimes experienced by the population aged 16 or over in England and Wales, as measured by the survey but grossed up to yield national figures. The bars at the bottom of the figure show crimes recorded by the police. For much of the thirty-year period the two trends have been consistent. However, the police statistics actually anticipated the 'real' fall in crime that began in 1995, showing a fall from 1993, arguably as a consequence of political

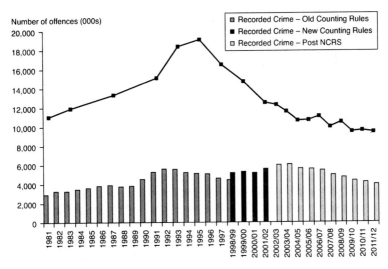

Figure 9.1 Crime trends in England and Wales, 1981–2012
Source: ONS (2012).

pressure on senior police to deliver reductions in crime. And secondly, the recorded crime statistics show a clear – but artefactual – increase in crime over the six years spanning the millennium, reflecting a succession of changes to the 'counting rules' issued by the Home Office. The fact that the trend in police statistics over this period was upward whilst the CSEW showed falls was exploited to the maximum by politicians. The opposition used the escalating police statistics as a political cosh with which to beat the Government[3] – even if anyone with any statistical literacy could see that this trend was misleading, and simply the result of substantial changes in recording practices.

It might be questioned whether this collection of surveys of victimisation can claim to be social *scientific* research. Much of the work has been a-theoretical, often presented as less methodologically problematic than it actually is; it has been much criticised on these grounds by academic criminologists, notably by Young (2011) whose very overstated, if witty, critique of the methods used by quantitative criminologists has been neatly undermined by Garland (2012). There are other limitations to the use of population surveys to measure crime. They are poor at measuring rare crimes, 'victimless' crimes and those with institutional rather than individual victims. They are good at capturing 'crimes of the poor' – burglary, car theft and street robbery – and bad at capturing 'crimes of the powerful' – environmental crimes and large-scale banking frauds, price-fixing cartels and so on. They may lag behind in the measurement of emerging crimes, such as internet fraud and other 'cybercrimes'.

Whatever their limitations however, their contribution has been significant. Countries with well-developed programmes of victim surveys now find themselves very much better positioned to understand crime trends than they were in the third quarter of the twentieth century. Without survey data on victimisation in England and Wales, for example, it would have been hard to move beyond speculation in discussing crime trends over the last three decades. With the CSEW, we have a reasonably accurate estimate of the extent of unreported crime, as least for offences against people and their personal property, and we can track shifts in patterns of reporting to the police and recording by the police.

The value of survey research of this sort is that they yield social indicators that constitute the building blocks for a better understanding about the drivers of crime. Thus it is clear that in the 1980s the police statistics overstated the rate of crime increase because the rate at which crimes were *reported to the police* rose – reflecting growth in phone ownership and in insurance cover. In the mid-1990s, it seems that the rate at which reported crime was *recorded by the police* fell – perhaps reflecting political pressure on the police to meet new crime targets. In the late 1990s and early 2000s recording rates grew again, and there is now increasingly clear evidence (ONS, 2014, p36) that recording rates have been falling off since the early 2000s, thus exaggerating the fall in crime. The CSEW findings since 2007 mean that we are quite well placed to say whether or not the global financial crisis and the associated period of recession in the UK had affected national crime rates. In the absence of a downward trend from the CSEW, the most obvious conclusion to draw would be that 'real' crime trends were forced upward by recession, but that financial cut-backs in police staffing – and in their capacity to record crime – over this period had simply served to mask the increase. The CSEW permits us to reject this – albeit sociologically persuasive – theory with some degree of confidence (see also van Dijk, 2013).[4] Crimes against individuals and their property have continued to fall throughout the financial crisis, even if there is some evidence that the police statistics have overstated the rate of this fall (ONS, 2013).

If crime surveys have improved academic and professional understanding of crime levels and trends, this understanding has failed to penetrate political and public debate about crime. Despite almost two decades of falling crime, the CSEW shows that majorities of the population think that nationally crime is still rising – even if majorities think that crime is falling in their neighbourhood. Until very recently neither the mass media nor politicians were prepared to accept that crime has been in decline. It would appear that pessimism about falling standards of behaviour is very deeply ingrained into the public and political consciousness. We shall return to this issue, and its implications, at the end of the chapter.

Police work and the impact of the police on crime

Crime surveys exploited developing survey technology to provide answers to questions that had historically been easy to ask but hard to answer – clearly occupying the Rumsfeld-ian territory of 'known unknowns'. By contrast, social research into policing has proved much more iconoclastic, revealing that policing institutions function in ways that are substantially at odds with received wisdom – charting 'unknown unknowns'. Until the 1960s the police in most developed countries were largely closed to independent or academic scrutiny, which enabled them to define their own role – or public perceptions of that role. This changed in the early 1960s in the US and the UK, and somewhat later in mainland Europe. In 1980 Ron Clarke and I characterised this body of research as undermining professional and popular assumptions about the police, which we called the 'rational deterrent' model of policing (Clarke and Hough, 1980: 2). The key assumptions of this model were that:

- the police were the primary agents of social control
- that social control and crime control were synonymous
- that police work was mainly to do with crime-fighting and the deployment of deterrent strategies.

More recently Reiner (2012) has described this process as a dialectical one. He identifies as the thesis the popular (and political/media) conception of the police as crime busters; the antithesis was formed by research showing that only a minority of police time was spent on crime, that the crime dealt with by the police was largely reported to them by the public, and that their deterrent impact was marginal. He describes the academic synthesis of these two positions as presenting the police as an emergency service with a capacity to deploy coercive force, whether to deal with crime or to resolve other problems that require immediate attention.

The sociological policing research of the 1960s and 1970s was genuinely path-breaking. Key pieces of work include Reiss's (1971) detailed observational research of police at work in the US, and at work in the UK by Banton (1964); both studies showed the wide range of demands, most of them non-criminal, made on the police. Punch's (1979) work characterised the police as a 'secret social service'. Egon Bittner's (1970, 1974) work remains very widely cited as providing an important redefinition of the police mandate; his two most quoted passages are probably his characterisation of police work as, 'something-that-ought-not-to-be-happening-and-about-which-somone-had-better-do-something-now!' (1974: 30), and his statement of the unique competence of the police:

> The specific capacity of the police is wholly defined in their capacity for decisive action…More specifically, that the feature of decisiveness derives from the authority to overpower opposition in the 'then and there' of the situation of action. *The policeman, and the policeman alone, is equipped, entitled, and required to deal with every exigency in which force may have to be used to meet it* (Bittner, 1974: 35).

Bittner's account of the police as an *emergency* service whose effective delivery relied on the capacity of deploying coercive force was consistent with accounts of what the police in developed countries actually do. Manning's (1977) influential account of policing in London confirmed the disjunction between the reality of day-to-day police work and the idea of policing as crime-busting that is embedded in political and media discussions of policing. He argued that the police were engaged in 'the dramatic management of the appearance of effectiveness' and that whatever effect they had on people's offending, this was mediated through symbolism – in other words, that an important dimension of policing operates at the symbolic level.

If these studies demonstrated that police work actually bore little resemblance to rational deterrent crime fighting, a series of important experimental US studies also showed that crime levels were unaffected by changes at the margin[5] in levels of car and foot patrol (see especially Kelling et al., 1974; Pate et al., 1986; and Clarke and Hough, 1985 for a summary). Yet further work suggested that the detective function was less one of systematic sifting of evidence that eventually identified the culprit, and more one of collecting straightforward evidence from victims and witnesses who had already identified the offender[6] (Greenwood et al., 1977).

Many of these studies from the third quarter of the twentieth century are now classics of social science research into policing, and remain as essential reference points for modern academic scholarship in the field. The argument that changing levels of police resources or policing strategies achieves at best small gains at the margin in terms of reduced crime remains broadly accepted (see also Bradford, 2011). The synthesis offered by Reiner (2012) stressing the role of the police as responding to emergencies would probably command widespread agreement from academic criminologists. It is significant, nevertheless, that most of the studies cited here were published in the 1970s and 1980s. Subsequent work has served to refine or develop sociological perspectives on the police function, but not to radically reshape the field. Indeed the political task facing academic criminologists working in this area has been to convey the same message to successive generations of politician – that the police function is complex, multi-faceted and less centrally to do with crime control than is popularly assumed.

Research into normative compliance in explaining conformity with the law

The third research theme to be considered here is work on the role of institutions in securing normative compliance (see also Bottoms, 2002). This can be seen as a natural development of research challenging an overly simple 'rational deterrent' conception of police function, in providing a fuller account of the processes by which compliance with authority is actually secured. Procedural justice theory has its roots in Weberian and Durkheimian sociology but emerged in the US over the last twenty-five years (Tyler, 2006, 2011a, 2011b; Tyler and Huo, 2002). It has tended to contrast instrumental and normative mechanisms for securing compliance, proposing that in many areas of behaviour, people's behaviour is guided by normative rather than instrumental considerations. The key propositions of procedural justice theory are that the institutions of justice can shape – to some extent – the norms that guide people's behaviour and that treating people fairly is the key to doing so. Trust and legitimacy are central concepts: it is proposed that fair treatment by those wielding authority builds trust; that trust confers legitimacy on the institution in question; and that if those who are subject to its authority confer legitimacy on it, they will comply with its requirements. Procedural justice theory has the – somewhat paradoxical – attraction of providing an instrumental justification for ensuring that the justice system acts with decency, fairness and legality.

The large body of US evidence is being increasingly supplemented by UK theorising and empirical work (e.g., Bottoms and Tankebe, 2012; Hough, Jackson and Bradford, 2013; Jackson, Bradford, Stanko et al.,

2012a; Jackson, Bradford, Hough et al., 2012b; Tankebe, 2013), to support procedural justice theory. Much of this relies on population surveys, which consistently demonstrate the expected correlations between fair treatment, trust, perceived legitimacy and compliance. There is rather less experimental research testing whether these relationships are in fact causal, though the work that has been done has been positive (notably Mazerolle, Antrobus, Bennett et al., 2013). Our own work (Hough and Sato, 2011; European Social Survey, 2011, 2012; Jackson et al., 2012b; Hough et al., 2013) has used the European Social Survey to examine variations across country; again we have replicated the hypothesised relationships between perceptions of fair treatment, trust, legitimacy and compliance, but we have also found that 'moral alignment' between institutions and those over whom they exercise authority is a critical legitimating factor, as proposed by Beetham (1991). That is, people are more likely to confer legitimacy on the police or the courts if they believe that these operate to the same moral values as themselves.

It is hard for someone immersed in a particular field of research to assess what levels of visibility and influence this body of work has achieved. In the US, procedural justice research appears to be well-established. Large numbers of papers on the subject are presented at the major criminological conferences and the ideas appear to be finding some traction amongst politicians and criminal justice managers. Part of the reason for this is that the sheer cost of the reliance of instrumental strategies involving mass incarceration is creating pressure to find more financially viable alternatives. Another factor may be the fact that procedural justice theory provides a useful and appropriate set of concepts to apply to the – increasingly popular – policing strategies that have a neighbourhood or community focus.

Research is less developed on this side of the Atlantic. There are various groupings of academics and police researchers that are active in the area. In the UK, in addition to our own work with the European Social Survey mentioned above, some significant work has been done within and for the Metropolitan Police Service in London (e.g., Jackson et al., 2012a) and the UK National Policing Improvement Agency, now the College of Policing (e.g., Myhill and Quinton, 2011, Myhill and Bradford, 2012). Various national bodies, such as the National Audit Office and Her Majesty's Inspectorate of Constabulary have made supportive comments or references to procedural justice. Civil servants within the Home Office and Ministry of Justice are familiar with the concepts. But it would be wrong to suggest that procedural justice ideas have achieved any real purchase on political or media discourse. Even though there has been cross-party support for neighbourhood policing (a variant of community policing that has been adopted nationally) politicians justify this not in terms of its legitimating capacity, but in terms of a partnership between police and public in 'the fight against crime'. It would seem that social scientific research can reach into the technocratic parts of the process by which justice policy is formulated but has much more difficulty in making itself heard amongst politicians. Let us now turn to the reasons for this.

Criminology and politics

In her first major speech after her appointment in 2010, Theresa May, the UK Home Secretary at the time of writing, told senior police officers, 'Your job is nothing more, and nothing less, than to cut crime'.[7] This sound-bite neatly encapsulates three key assumptions that the criminological research discussed here has called into question: that crime is rising, that police work is solely about crime control, and that the police have the capacity to drive crime down. This is not intended as a partisan criticism of the current UK coalition government. Politicians from all the main political parties have struck similar postures over the last two decades. In the UK (or at least in Westminster politics) talking tough on crime is routine for Home Secretaries, Justice Secretaries, Prime Ministers and their opposition shadows.

The main reasons for this are to be found in the rapid increase in the temperature of the criminal justice debate, which can be dated to the early 1990s.[8] The main heating source was provided by year-on-year increases in recorded crime – averaging 6% per year in England and Wales for over three decades, which not surprisingly was reflected in growing public concern about crime. But a more proximate reason was that New Labour was overhauling policy in many key areas, including crime, in anticipation of the coming election. These developments created the preconditions for the then shadow Home Secretary, Tony Blair, to mount an effective challenge to the Conservative Party's status as 'the party of law and order'.[9] He famously promised to be 'tough on crime, tough on the causes of crime', initially in early 1993.[10] The media construed the emphasis to be on the first half of the promise, which created considerable pressure on the Government to show similar steel, and Michael Howard, the Home Secretary, responded later in the year with his own sound-bite, 'Prison works'. Since then, the competition between the parties to 'out-tough' each other has been relentless. The upshot has been a process of over-simplification of the issues in political (and media) discussion of crime – and there has been little patience for academics who insist that things aren't that simple.

This process of 'politicisation' and over-simplification has been amplified by two further factors. The first of these is the diminished role of the 'technocratic expert' in social policy, and a greater responsiveness to the voice of the public (see also Giddens, 1991). This trend has been particularly marked in countries such as the UK and the US with adversarial political systems, on the one hand, and attachment to neo-liberal market principles, on the other. In these countries, political responsiveness to public opinion has taken on an overtly populist quality; in criminal justice, the phenomenon of penal populism is well documented, whereby political leaders promote policies largely or entirely for the electoral advantage they confer, rather than from knowledge or conviction that they are the best policies (e.g., Roberts, Stalans, Indermaur et al., 2003).

Intertwined with this has been the development of a 'small state' style of governance in which politicians specify the *outcomes* required of state institutions such as the police, usually in the form of numerical targets, but leave the detail of the *processes* to local agencies. These principles of 'New Public Management' (NPM) are often applied in parallel with processes of 'marketisation' – where private sector companies compete for contracts to provide public services – and consumer choice – where service recipients can exercise control over the services they receive (see also Hood, 1991).

I have argued elsewhere (e.g., Hough, 2007) that criminal justice policy has suffered badly from the combined effects of penal populist and NPM policies. Once politicians adopt a crude and simple instrumental discourse about 'the war on crime', they find themselves trapped within its logic – partly because they judge that this is the only one that will be favourably received by the media (and, behind them, the electorate); and equally, the logic of NPM has driven politicians to adopt simple numerical targets that are built around crime and detection statistics. In combination, the increased weight given by politicians to the public voice and the increased uptake of forms of governance through numerical target-setting have squeezed out subtlety from political discourse about crime and justice.

Whether these trends towards over-simplification of policy and insulation of policy from academic research are structural – in the sense of being an inherent feature of politics in late-modern industrialised societies – remains to be seen. One school of thought is that the 'punitive turn' that is a consequence of penal populism is restricted to specific countries with particular political traditions and patterns of media ownership. Thus Tonry (2004) has argued that the US – and to a lesser extent the UK – are the outliers, and that politicians in mainland Europe can and should resist the pressures that draw them away from rational policy. On the other hand, plenty of mainland-European criminologists argue that the US and UK are less outliers than the advance guard of a new and nastier form of adversarial, media-led politics (e.g., Sack and Schlepper, 2013).

There is room for a little optimism. In the UK at least, there has been a retreat from the worst excesses of NPM, as well as frequent calls for more mature and less adversarial forms of politics. And the falls in the main indices of crime across several jurisdictions may have the effect of cooling the climate of criminal and penal policy debate. Certainly polls measuring public anxiety about social issues have shown a reduction in crime concerns – which appear displaced by concerns about the economy. And following a series of media scandals and the Leveson Inquiry (Leveson, 2012), there are signs that politicians are prepared to be more robust in their handling of the press. It is as yet unclear as to whether these developments will create more political space for engagement with social scientific research.

There are probably things that academic criminologists can do to increase the chances of more fruitful engagement with politicians. In the first place, this requires more positive enthusiasm for engaging in 'public criminology', with a view to improving what Loader and Sparks describes as taking on the role of a 'democratic under-labourer':

> Democratic under-labouring is: committed to both participating within, and to facilitating and extending, institutional spaces that supplement representative politics with inclusive public deliberation about crime and justice matters, whether locally, nationally or in emergent transnational spaces. In this regard, the public value of democratic under-labouring lies not in cooling down controversies about crime and social responses to it, but in playing its part in figuring out ways to bring the heat within practices of democratic governance.... If one was to encapsulate all the above in a single phrase it would be this: intellectual ambition, political humility.
> (Loader and Sparks, 2010: 132)

It is questionable whether the majority of academic criminologists are ready to grasp this role enthusiastically. This is because engaging with politicians and their policy officials is quite a time-consuming process, involving the careful building of networks and relationships. which can often be distracting from what academics might reasonably see as their 'day jobs' – teaching and publishing academic works. Moreover, reflecting attitudes in the wider public, academia has its fair share of cynicism about politicians. However, the incentives that are now being built into the UK funding system for higher education include rewards for research that can demonstrate 'impact' on policy or on broader social wellbeing. This may focus minds on contributing not only to the body of academic knowledge but also to the social good, measured in the short or medium term by contributions to the political process.

However, nurturing an ambition to contribute to the policy process is not the same as achieving it. Several other preconditions are required. Considerable planning and positioning is usually needed by any policy researcher to ensure that their research gets read by the right people, and there is also an element of luck and happenstance in the process. Having something coherent to say is, of course, the first requirement. To be able to say it with authority is also important, and building authority in the eyes of the right people is a slow process. Timing is a critical factor, as politicians need ideas at different points in the political cycle – notably when in opposition and developing a new set of policies for the election. Scale can be important, as people tend to set most store by large-scale research. And having non-academic allies – or, at least, sympathetic listeners – is critically important, whether these are politicians and their advisers, civil servants, think-tanks and lobbying groups, criminal justice agencies or journalists. What is undeniably the case – and what is very obvious to anyone who has engaged with policy for any length of time – is that criminologists are indeed minor players with small voices in the policy arena, and that their research will achieve little if they fail to foster, in some way or other, forms of reach into the political process additional to the publication process.

Notes

1. In its early years the discipline had a symbiotic relationship with Home Office research and policy, and indeed it was this closeness that prompted a reaction that involved a more critical stance exemplified by the establishment of the National Deviancy Conference in 1968, which has dominated sociological criminology ever since.
2. Adolphe Quetelet, the Belgian sociologist and statistician – see Zauberman and Robert (2011).
3. The new counting rules had an especially inflationary effect on violent crimes (see also Hough and Maxfield, 2007).

4. The downward trend in crime – exhibited by many developed countries – stands in need of explanation, and criminologists have not (yet) done a good job here. Improved security and better anti-theft design is clearly part of the story. To some extent the growth in crimes poorly measured by both police statistics and surveys will have offset the falls in conventional 'volume' crimes like burglary and vehicle crime, but the new forms of crime almost certainly involve different victims and offender groups than the traditional ones.
5. Few would argue that there is no impact in *gross changes* in police levels, for example when saturation patrolling is introduced, or *all* police presence is removed as in the case of police strikes.
6. This is not to suggest that detections *never* result from careful sifting of forensic evidence, and techniques such as DNA testing make such cases more frequent. But they did not – and very probably still do not – represent the typical route to detection.
7. www.gov.uk/government/speeches/police-reform-theresa-mays-speech-to-the-national-policing-conference
8. A metaphor neatly built upon by Loader and Sparks (2010), who explore the scope of various 'cooling devices'.
9. Downes and Morgan (2007) suggested that the Conservative Party started this process in 1979, making support for the police a key political issue. However, the Labour Party did little to challenge the government's crime policies until 1992.
10. Interview on BBC Radio 4 'The World This Weekend', 10 January 1993. See also Blair (1993).

Bibliography

Banton, M. (1964) *The Policeman in the Community*. London: Tavistock.
Beetham, D. (1991). *The Legitimation of Power*, London: Macmillan.
Bittner, E. (1970), *The Functions of Police in Modern Society*, Chevy Chase, MD: National Institute of Mental Health.
Bittner, E. (1974) 'Florence Nightingale in pursuit of Willie Sutton': a theory of the police.' In H. Jacob (ed.) *The Potential of Reform of Criminal Justice*. Beverly Hills, Cal.: Sage.
Blair, T. (1993). Why crime is a socialist issue. *New Statesman and Society*, 29 January, 27–28.
Bottoms, A. (2002). Compliance and community penalties. In A. Bottoms, L. Gelsthorpe and S. Rex (eds), *Community Penalties: Change and Challenges*, Cullompton: Willan, 87–116.
Bottoms, A., and Tankebe, J. (2012). Beyond procedural justice: a dialogic approach to legitimacy in criminal justice. *Journal of Criminal Law and Criminology*, 119–170.
Bradford, B. (2011). *Police Numbers and Crime Rates – A Rapid Evidence Review*, London: Her Majesty's Inspectorate of Constabulary.
Clarke, R. and Hough, M. (1980). *The Effectiveness of Policing*, Farnborough: Gower.
Clarke, R.V.G. and Hough, J.M. (1984). *Crime and police effectiveness*. Home Office Research Study No. 79. London: HMSO.

Downes, D. and Morgan, R. (2007). No turning back: the politics of law and order into the millennium. In M. Maguire, R. Morgan and R. Reiner (eds), *The Oxford Handbook of Criminology* (4th edn). Oxford: Oxford University Press.

European Social Survey (2011). Trust in justice: topline findings from the European Social Survey. *ESS Topline Results Series Issue 1*, by J. Jackson, M. Hough, B. Bradford, T. M. Pooler, K. Hohl and J. Kuha.

European Social Survey (2012). Policing by consent: understanding the dynamics of police power and legitimacy. *ESS Country Specific Topline Results Series Issue 1 (UK)*, by J. Jackson, M. Hough, B. Bradford, T. M. Pooler, K. Hohl and J. Kuha.

Garland, D. (2012). Criminology, culture, critique: a review of Jock Young, the criminological imagination. *British Journal of Criminology*, 52(2), 417–425.

Giddens, A. (1991). *The Consequences of Modernity*, Cambridge: Polity Press.

Greenwood, P. W., Petersilia, J. and Chaiken, J. (1977) *The Rand Criminal Investigation Study*. Lexington, MA: D. C. Heath.

Home Office (1896). *Criminal Statistics for England and Wales 1894*, C 8072.

Hood, C. (1991). A public management for all seasons. *Public Administration*, 69(1), 3–19.

Hough, M (2007). Policing, new public management and legitimac. In T. Tyler (ed.), *Legitimacy and Criminal Justice*, New York: Russell Sage Foundation.

Hough, M. and Maxfield, M. (eds) (2007) *Surveying Crime in the 21st Century*. Cullompton: Willan Publishing.

Hough, M. and Mayhew, P. (1983). *The British Crime Survey: first report*. Home Office Research Study No. 76. London: HMSO.

Hough, M. and Sato, M. (eds) (2011). Trust in justice: why it is important for criminal policy, and how it can be measured, Final report of the Euro-Justis project. Helsinki: HEUNI. Available at http://www.icpr.org.uk/media/31613/Final%20Euro-Justis%20report.pdf accessed 24 September 2014.

Hough, M., Jackson, J. and Bradford, B. (2013). Trust in justice and the legitimacy of legal authorities: topline findings from a European comparative study. In S. Body-Gendrot, M. Hough, R. Levy, K. Kerezsi and S. Snacken, *European Handbook of Criminology*, London: Routledge.

Hough, M., Jackson, J., Bradford, B., Myhill, A. and Quinton, P. (2010). Procedural justice, trust and institutional legitimacy. *Policing: A Journal of Policy and Practice*, 4(3), 203–210.

Jackson, J., Bradford, B., Hough, M., Kuha, J., Stares, S. R., Widdop, S., Fitzgerald, R., Yordanova, M. and Galev, T. (2011). Developing European indicators of trust in justice, *European Journal of Criminology*, 8, 267–285.

Jackson, J., Bradford, B., Stanko, E. A. and Hohl, K. (2012a). *Just Authority? Trust in the Police in England and Wales*, London: Routledge.

Jackson, J., Bradford, B., Hough, M., Myhill, A., Quinton, P. and Tyler, T. R. (2012b). Why do people comply with the law? Legitimacy and the influence of legal institutions. *British Journal of Criminology*, 52(6), 1051–1071.

Kelling, G. L., Pate, A. M.,Dieckman, D. and Brown, C. (1974). *The Kansas City Preventive Patrol Experiment*. Washington, DC: Police Foundation.

Kitsuse, J. I. and Cicourel, A. V. (1963). A note on the use of official statistics. *Social Problems*, 1(2), 131–139.

Leveson, B. (2012). *An Inquiry Into the Culture, Practices and Ethics of the Press*, London: The Stationery Office.

Loader, I. and Sparks, R. (2010). *Public Criminology*, London: Routledge.

Manning, P. (1977). *Police Work: the social organisation of policing.* Cambridge, Mass: MIT Press.

Mazerolle, L., Antrobus, E., Bennett, S. and Tyler, T. (2013). 'Shaping citizen perceptions of police legitimacy: a randomised field trial of procedural justice'. *Criminology*, 51(1), 33–63.

Morris, R. M. (2001). Lies, damned lies and criminal statistics: reinterpreting the criminal statistics in England and Wales. *Crime, History & Societies*, 5(1), 111–127. Available at http://chs.revues.org/index784.html, accessed 24 October 2014.

Myhill, A. and Bradford, B. (2012). Can police enhance public confidence by improving quality of service? Results from two surveys in England and Wales. *Policing and Society*, 22(4), 397–425.

Myhill, A. and Quinton, P. (2011). *It's a Fair Cop: Police Legitimacy, Public Cooperation, and Crime Reduction: An Interpretative Evidence Commentary*, London: National Policing Improvement Agency (now College of Policing). Available at http://www.college.police.uk/en/docs/Fair_cop_Full_Report.pdf, accessed 19 May 2013.

ONS (2012). Crime in England and Wales – Quarterly First Release, March 2012. Available at www.ons.gov.uk/ons/rel/crime-stats/crime-statistics/period-ending-march-2012/stb-crime-stats-end-march-2012.html accessed 24 September 2014.

ONS (2013). *Relationship between Crime Trends Measured by the Crime Survey for England and Wales and the Police Recorded Crime Series*, London: Office for National Statistics.

Pate, A. M, Lavrakas, P. J., Wycoff, M. A., Skogan, W. G., and Sherman, L. W. (1986). *Reducing fear of crime in Houston and Newark: A summary report.* Washington, DC: Police Foundation.

President's Commission (1967). The challenge of crime in a free society: a report by the president's commission on law enforcement and administration of justice. Washington: US Government Printing Office. Available at https://www.ncjrs.gov/pdffiles1/nij/42.pdf, accessed 16 April 2013.

Punch, M. (1979) 'The secret social service', in S. Holdaway (ed.) *The British Police.* London: Edward Arnold.

Reiner, R. (2007). *Law and Order: An Honest Citizen's Guide to Crime Control*, Cambridge: Polity Press.

Reiner, R. (2010). *The Politics of the Police* (4th edn). Oxford: Oxford University Press.

Reiner, R. (2012). *In Praise of Fire-Brigade Policing: Contra Common Sense Perceptions of the Police Role*, London: Howard League for Penal Reform.

Reiss, A. J. (1971) *The Police and the Public. New Haven*, Conn: Yale University Press.

Roberts, J. V., Stalans, L. S., Indermaur, D. and Hough, M. (2003). *Penal Populism and Public Opinion. Findings from Five Countries*, New York: Oxford University Press.

Sack, F. and Schlepper, C. (2013). Changing definitions of the criminal law in Germany in late modernity. In S. Body-Gendrot, R. Lévy, M. Hough, S. Snacken and K. Kerezsi (eds), *The Routledge Handbook of European Criminology*, London: Routledge.

Skogan, W. (2006). Asymmetry in the impact of encounters with the police. *Policing and Society*, 16, 99–126.

Sparks, R., Genn, H., and Dodd, D. (1977). *Surveying victims*. London: Wiley.

Sunshine, J. and Tyler, T. (2003). The role of procedural justice and legitimacy in public support for policing. *Law and Society Review*, 37(3), 513–548.

Tankebe, J. (2013). Viewing things differently: the dimensions of public perceptions of legitimacy. *Criminology*, 51(1), 103–135.

Tonry, M. (2004). *Punishment and Politics: Evidence and Emulation in the Making of English Crime Control Policy*, Cullompton: Willan Publishing.

Tyler, T. R. (2006). *Why People Obey the Law*, Princeton: Princeton University Press.

Tyler, T. R. (2011a). *Why People Cooperate: The Role of Social Motivations*, Princeton: Princeton University Press.

Tyler, T. R. (2011b). Trust and legitimacy: policing in the USA and Europe. *European Journal of Criminology*, 8, 254–266.

Tyler, T. R. and Huo, Y. J. (2002). *Trust in the Law: Encouraging Public Cooperation with the Police and Courts*, New York: Russell Sage Foundation.

van Dijk, J., van Kesteren, J. and Smit, P. (2007). *Criminal Victimisation in International Perspective: Key Findings from the 2004–2005 ICVS and EU ICS*, The Hague: WODC (RDC) Ministry of Justice.

van Dijk (2013). 'It is not just the economy': explaining post–World War II crime trends in the western world. In S. Body-Gendrot, M. Hough, R. Levy, K. Kerezsi and S. Snacken (eds), *European Handbook of Criminology*. London: Routledge.

Wikström, P-O. H., Oberwittler, D., Trieber, K. and Hardie, B. (2012). *Breaking Rules: The Social and Situational Dynamics of Young People's Crime*, Oxford: Oxford University Press.

Young, J. (2011). *The Criminological Imagination*, Cambridge: Polity Press.

Zauberman, R. and Robert, P. (2011). Sur l'évolution de la statistique criminelle et plus largement de la mesure de la délinquance. *Electronic Journal for History of Probability and Statistics*, 7(1), June, 1–13. Available at http://hal.archives-ouvertes.fr/docs/00/74/31/38/PDF/2011-PhR-RZ-A_volution_statistique_criminelle_JEHPS.pdf, accessed 16 April 2013.

10
Understanding the Arab Spring

Stuart Croft and Oz Hassan

Introduction

Social Science has made extraordinary contributions to a range of fields that impact directly on the lives of states and their citizens, in fields such as economics and society. What is sometimes less obvious, though, is that social science also has a major role to play in understanding the nature of international security. Often the loudest voices in understanding violence around the world belong to the journalist, with the validity of being 'on the spot' and 'in the moment'. Of course, such reporting is vital. But the downside of, in particular, 'in the moment' is that often a single narrative is reported, becoming a way of understanding a series of complex events through a single prism. It is the task of social science to be able to take that singular narrative and to be able to place it in a wider context, to understand the powerful impact of history and of culture, of faith, of gender and of constructions of ethnicity, and of wider power balances.

This chapter seeks to do precisely that; to show how a significant event – here the 'Arab Spring' – can be dominated by a single narrative of cause and effect. We then show how a social science that is engaged with understanding wider and deeper causes can throw further light onto such complexity.

The prevailing narrative of the Arab Spring

The prevailing narrative of the Arab Spring[1] starts on 17 December 2010. It argues that Mohammed Bouazizi, a twenty six-year-old Tunisian, set himself alight outside the Sidi Bouzid regional council house, rejecting authoritarianism and providing a rallying cry for disaffected youth. It is argued that this provided the impetus, combined with the wikileaks release of classified American cables on the Tunisian regime, for 'solidarity' uprisings spreading throughout Tunisia in December 2010. This culminated, by 14 January 2011, in Tunisian President Ben Ali and his family fleeing to Saudi Arabia, ending Ben Ali's twenty-three-year dictatorial rule in just twenty-eight days. The Tunisian people had succeeded in overthrowing their authoritarian regime and opened up a new realm of political possibility. More widely, Tunisia had demonstrated to the peoples of the Middle East and North Africa (MENA) region that change is possible and that autocrats can be overthrown through popular protest. A single act of existential despair and a demonstration of public defiance rapidly and unexpectedly escalated into the Arab Spring.

The narrative continues by highlighting events in Egypt, as 25 January 2011 became the 'Day of Rage' sparking anti-government demonstrations across the country. In Tahrir Square, Cairo, youth movements gathered to call for President Hosni Mubarak to leave, whilst the government responded by blocking Facebook and Twitter, before finally shutting down mobile phone and internet services. The police met protestors with violence, driving vehicles into crowds, throwing tear gas and firing shotguns filled with metal pellets at random. Protestors began setting fire to buildings, targeting the police and defying the imposed curfew. By 11 February, as protestors began marching towards the presidential palace, President Mubarak finally heeded calls to step down, ending his nearly thirty years of autocratic rule. Within less than two months the status quo in the MENA had been irreparably changed, and popular protests began to spread to greater and lesser degrees in Algeria, Bahrain, Iran, Iraq, Jordan, Kuwait, Lebanon, Morocco, Oman, Pakistan, Saudi Arabia and Yemen. In Libya, a civil war against Colonel Gaddafi's regime began, and threats of genocide led the United Nations Security Council passing Resolution 1973, to back a NATO intervention and to secure a no-fly zone. Whilst in Syria, March 2011 saw the beginnings of an uprising that would become increasingly violent and ferment into a longer civil war.

On the face of it, this narrative represents the Arab Spring as a positive move for the peoples of the MENA region. It demonstrates the opening of political spaces and a possible democratic future, which could result in a greater appreciation of human rights, less violent conflict, and the dividends of a democratic peace. It highlights, therefore, that the importance of the Arab Spring is global, providing the possibility of a greater level of international security and stability. With the MENA being centrally linked to global energy and trade systems, a stable and secure region is important beyond Arab capitals. Moreover, having partners in the region that cooperate on issues of counter-terrorism and counter-proliferation is important to national security concerns of Western governments. However, it is far from clear that the Arab Spring will lead the world to a more stable environment. Since the uprisings, there have been significant tensions over issues of transitions to civilian rule, exacerbated ethnic and religious conflicts, raised tensions between Israel and its neighbours, emboldened terrorist groups targeting Western workers and officials, renewed fears that Iran will become a more powerful regional actor, and fear that Islamist parties will undermine the transition processes underway. This raises questions over whether the fall of autocratic regimes will lead to turbulent transitions to stable democracies, the emergence of renewed autocratic rule, or the creation of failed states? Similarly, questions over the exact geopolitical orientation of the region are being raised, as it is not clear that the emerging governments can, or would, cooperate fully with the pursuit of Western interests in the region. The Arab Spring has therefore introduced both hope and considerable uncertainty into the international system for years to come.

Given the profound nature and implications of these events, it is fundamental that the Arab Spring be fully understood, beyond that of the prevailing narrative. It is only through fully understanding the complex reasons for these events, and the interconnected implications of them, that informed policies can be made. Within this context, there are many excellent commentaries available from media and think tank organisations that provide insight into these events, and which go some way towards helping develop our understanding. Such sources provide a rapid and necessary overview that is both informative and high impact. However, what distinguishes such sources from the social sciences is that the latter seeks to systematically combine theoretically informed methodological insights and corresponding research methods into the analysis to derive research-based conclusions. That is to say that the social sciences work within the frameworks derived from a long historical tradition of the philosophy of social sciences, so as to produce a rigorous and accessible body of knowledge that helps explain/understand social phenomena. When a social scientist seeks to illuminate the issues around the Arab Spring, therefore, the aim is to provide an analysis informed through applied research methods, which helps identify and prioritise factors contributing to highly complex issues, which in turn narrows the level of uncertainty. This, in itself, is a valuable enterprise for furthering the social sciences and informing the academic debate, but importantly this also allows the social sciences to inform multiple stakeholders ranging from the broader public to policy makers. This is not to suggest that social scientists are homogeneous in their accounts, but rather that they provide a series of broader contextualisations from which to inform, making highly complex and fast moving situations more accessible.

For social scientists, it is not possible to reduce the reasons for the Arab Spring to single events. Whilst it is clear that the self-immolation of Mohammed Bouazizi is important, the philosophy of social sciences teaches us that social phenomena occur within multifaceted contexts. There is of course disagreement over how to identify and estimate the importance of factors within this context, which are determined by particular methodological schools' ontological and epistemological positions. Whilst some schools of thought seek to explain events parsimoniously, others look for more complex human reasons for action to understand why events occur. There are also particular weightings made to structural factors compared to the role of agents; the role of material phenomena compared to the role of ideas or discourses; and different emphases made on how to understand the persistence of political continuity and emergence of political change. There are also debates about whether quantitative, qualitative or a mix of both methods should be used in the research strategy. These debates underlie social scientific research, and whilst these philosophical and metatheoretical debates are in and of themselves very complex and nuanced, when applied to an event such as the Arab Spring they help broaden out the analysis and raise awareness of the issues. They help identify what is, and what is not important, within the context of reasoned parameters. Thus, what we offer below is a multilevel approach to understanding the Arab Spring, that explores these debates at the individual, state and international levels, in an attempt to shed light on why these events occurred and what implications they have for Western policy.

Demographics, technology and pluralism

A distinctive feature of the Arab Spring has been the central involvement of youth movements across the region. This can be attributed to regional demographics, where the under thirty population is the majority. In 2010, 51% of the Tunisian population was under thirty, whilst in Egypt and Libya it was 61%, Syria 66% and as high as 73% in Yemen (US Census Bureau, 2010). Indeed, one third of the overall MENA population is between the ages of fifteen and twenty-nine, compared to 20% in the United States and 18% in Europe (Feiler, 2011: 113). For some, this 'youth bulge' provided the structural impetus for unrest. This, it is argued, is comparable to the youth protests and unrest in the US and Europe in the 1960s, driven by the post–Second World War baby boom. Indeed, some commentators go so far as to argue that knowledge of this demographic bulge should have provided the ability to predict a 'population bomb' in the region in advance of the Arab Spring. However, demographic information does little in helping to understand why youth movements rose up. It is a structural factor to consider within the analysis, rather than a reason for action. As Ragui Assaad argues, 'demographics have played an important role, not because they are the problems themselves but because they have exacerbated other serious problems that youth are having ... having large numbers of people who are very frustrated at their inability to turn their education into productive jobs, has really exacerbated the problems' (LaGraffe, 2012).

For many social scientists, the prevalence of poor economic conditions in the region has become central to understanding the Arab Spring. The MENA region has the highest youth unemployment figures in the world at 24% in 2009. This is more than double the adult unemployment rate in these countries, which is approximately 10%. Moreover, this is even more problematic given that the MENA has the world's lowest labour force participation rates, as a result of women not entering the workplace and many male and female youths leaving, and being put off of entering, the work force (Roudi, 2011). Thus, whilst the youth population represents significant economic potential, chronic underfunding in education and lack of employment opportunities has created poverty, inequality, large-scale unemployment and a loss of individual dignity. Thus, it is demographics combined with wider discontent regarding Arab political economies that is increasingly seen as the necessary conditions for inspiring the revolutions.

The combination of large-scale youth unemployment and alienation was certainly evident in the spring of 2008, when large-scale popular dissent originally bubbled to the surface around the MENA. In Egypt, textile workers in the town of Al-Mahalla Al-Kobra took to the streets to complain about increasing inflation undermining their already low pay. Associated supporters of this group would later be instrumental members of the 6 April movement, which was central to the 2011 struggles in Tahrir Square. In Tunisia, January 2008 saw an early rendition of the small-scale protests, similar to those that would escalate into a full revolution less than three years later. In Morocco, protests forced the government to cancel intended bread subsidy reductions. Small-scale demonstrations also took place in Algeria, Lebanon, Jordan and Yemen as, 'the cost of basic foodstuffs skyrocketed thanks to a combination of high oil prices, poor harvests, rising speculative investment in commodities, and biofuel subsidies that discouraged farmers from growing food' (Noueihed and Warren, 2012: 24). Within such a context, the initial disquiet in 2008 was a fundamental stepping-stone towards the 2011 revolutions.

The importance of the social unrest in 2008 is twofold. Firstly, it demonstrates that large-scale youth unemployment and alienation, even when leading to protests, is not enough to guarantee the onset of a revolution. Secondly, 2008 provides a point of comparison with the 2011 revolutions. What's apparent about events in 2008 is that many of the youth movements in the region reflected on this period to inform their future actions and develop new revolutionary strategies. These events inspired the youth to better communicate and mobilise. That is to say that the 2008 uprisings provided lessons for these groups to learn from. For example, it was during the 2008 uprisings that the April 6 movement was formed, to report on the Al-Mahalla Al-Kobra strikes. As part of this reporting strategy they drew on what were then new social media platforms, such as Facebook, Twitter, Flickr, Blackberry Messenger, and blogs. This was to allow them to be 'citizen journalists', enabled by the tools of the new web 2.0 environment. These were the same tools they deployed in 2011, after years constructing their networks, which proved to be 'a crucial platform for both organisation and real-time report from the street [during the Egyptian revolution]. Protesters not only knew what was happening a few blocks away or across town, they were also tracking in real time, what was happening across Egypt' (Khalil, 2011: 148). In addition to social media therefore, the proliferation of smart phones adds a hardware dimension to the analysis. However, this should not overshadow the importance of phone calls and texts during the revolution, especially once social media sites were shut down. Similarly, it is clear from multiple analyses that satellite television, and in particular 24 Hour News channels such as Al Jazeera and Al-Arabiyya played a crucial role in providing alternative sources of news to government-controlled media outlets. What unites these three communications methods, therefore, is not their 'novelty', but the manner in which they pluralise communication and how they were used as *sources* of information and *tools* for mobilisation; both of which challenged the power and legitimacy of authoritarian regimes.

The importance of 'virtual space' therefore is not viewed as a determinant factor for the revolutions in and of itself, but viewed as a tool for agents working within authoritarian structures. The ability for mass communication not only allowed youth social movements to correspond and prepare for the revolutions inside their countries, but also to outline best practice across the region, and across the globe. For example, when interviewing leading members of the 6 April movement, they made clear that they studied 'the American civil rights movement' and in particular 'Martin Luther King, Rosa Park and Malcolm X'. They also studied the struggles in 'Poland, Georgia, Ukraine and Serbia', which led them to develop relations with Otpor! in Belgrade. This led the 6 April movement to adopt the same clenched fist symbol as Otpor!, which was so visible in Tahrir Square during the revolution (Hassan, 2011, 2013). In addition to learning lessons from around the globe, they were also able to gain global recognition and support via their social media platforms. With authoritarian regimes demographic being considerably older than these youth movements, ultimately they failed to spot the implications and importance of these new social tools until it was too late. At the level of individual citizens a mixture of demographics, alienation, communication, learning, the utility of technology, and the pluralisation of information fostered a wider revolutionary consciousness. This complex mixture allowed individual agents to collectively challenge their respective regimes with extraordinary effectiveness.

Political economy and military decisions

Beyond the level of individual agents, there are also structural issues behind the Arab Spring, at the state level. For example, a distinguishing feature of the Arab Spring has been how throughout the MENA, it is one party presidential systems that have been successfully challenged from below. That is to say, that there were/are structural differences between the authoritarian regimes in Tunisia, Egypt, Libya, Syria and Yemen, compared to the Gulf Monarchies best represented by Saudi Arabia. Notably, however, the latter do not believe that they are immune from the 'demonstration effect' that has led to the spread of the revolutions. They have, however, been able to draw on their considerable oil wealth to help prevent poor economic conditions to help prevent poor economic conditions from enabling the fermentation a broader revolutionary consciousness. For example, in the aftermath of the initial revolutions Saudi Arabia announced a massive government subsidy package which added $130 billion in spending projections over the next five years. Monarchies have also proven more flexible by making moves to devolve power down towards other institutional bodies, and have been able to make stronger claims to religious-political legitimacy. These were not options available to the one party presidential systems that emerged as former colonial powers withdrew from the region in the post–Second World War period.

Indeed, one party presidential system has long based their legitimacy on constitutional propriety, even as authoritarian leaders constructed 'security states' (Owen, 2012). These strongmen committed despotic acts against their own populace, whilst believing themselves sovereign unifiers of their republics. To 'save' their people they imprisoned, tortured, and executed their enemies. This allowed regimes to cement their power for the all too numerous years they thought were necessary to lift their populations through economic and social development. What emerged were systems that gave greater freedoms to the few, whilst withdrawing social, political and economic freedoms of the many. This concentration of power within the state was made possible by institutionalising particular structures. As Roger Owen details,

> At the apex of such systems stand the presidential office, the presidential family, and a small group of advisers drawn from the military, the security services, and the business elite. Next in order of importance come the senior members of the army, the intelligence agencies, and the police, together with a small group of crony capitalists who obtain access and influence in exchange for a role in providing additional resources for the regime in terms of money and sometimes, organisational skills. Under them are the agencies of civilian administration, the ministries, and the provincial governors, as well as the most important centres of ideological control: the educational apparatus, the official media, the tame judiciary, and the equally tame religious establishment. (Owen, 2012: 38)

Understanding these structures is paramount to a deeper appreciation of the Arab Spring, because they reveal a lack of political and economic transparency combined with high levels of institutional corruption. To maintain such systems friends of these regimes required rewards. This was evident in tax systems that favoured the few, penalised the many, and exacerbated wider poverty throughout MENA societies. Government functioning for the few led to the concentration of wealth amongst shrinking middle classes, making economic inequality all the more evident. Within these state structures, even where there was economic growth, as states moved from renter models to greater economic liberalisation, this was ineffective. For example in spite of Tunisia and Egypt, prior to 2010, seeming to perform well in terms of economic growth, this did not translate into jobs. The importance of this cannot be over-stated, because moves towards liberalising their economies were part of a regime-driven survival strategy. Regimes believed that economic growth would prevent the onset of revolutions by diffusing popular dissatisfaction and helping to avoid legitimation crises. Such policies ultimately proved ineffective because this was jobless growth. Accordingly, neoliberal economic policies that favoured privatisation and open markets failed to deliver 'trickled down' wealth, but did succeed at growing resentment (Hassan 2011).[2] Ultimately, the survival strategy adopted by regimes relied on flawed economic assumptions and further alienated large sections of their populations.

In addition to structural issues with Middle Eastern political economies, the structural role of the militaries also proved crucial to successful revolutions in the Arab Spring. Arab militaries, since the coups in the 1950s and 1960s, have long been significant institutions and arbiters of power in their respective societies. What was, however, highly noticeable during the unfolding of events was that militaries in each country behaved differently. For example, in Tunisia and Egypt, their militaries chose not to use force against the demonstrators. This removed a structural impediment faced by protesters, and ultimately allowed El Abidine Ben Ali and Hosni Mubarak to be removed from power. Indeed, when Ben Ali ordered the military to fire upon protesters, they refused. This allowed them to emerge as a stronger political institution after his removal, declaring that the military was the 'guarantor of the revolution'. Similarly, in Egypt, the military decided not to intervene against the revolts. However, they did make a more directly political decision to force Mubarak to step down, which reflected their growing disquiet with being sidelined and concerns with attempts to transfer power to Mubarak's son Gamel.

The alternative military model was demonstrated in Libya and Syria. In contrast to Egypt and Tunisia, the Libyan and Syrian militaries were not autonomous institutions. They were a fundamental part of the ruling regimes. Indeed Colonel Gaddafi's son headed the most important military units involved in early attempts to suppress the revolt. In Syria, the al-Assad family and members of their al-Matawirah tribe dominated the upper ranks of the military, intelligence and security services with the sole purpose of protecting the family and the minority Alawite religious sect. These militaries were therefore the hard power institutions of the regimes themselves, embedded within their state structures. The survival of their regimes was explicitly linked to the survival of their militaries. As such, it is unsurprising that these were the revolts that became violent, as the militaries chose to intervene in the revolts. However, whilst in Libya rebels were able to defeat Gaddafi, with the help of NATO, an ongoing civil war has broken out in Syria. Within such a context it is clear that the institutional distribution of power within the state is important to understanding the Arab Spring. Whilst too much economic power concentrated amongst regime elites and crony capitalists fermented legitimation crises, how military power is institutionalised within the state matters to the outcome of the revolutions themselves. Moreover, as the case of Libya directly shows, these are not separate from the global context.

Historical contexts and global effects

Understanding the Arab Spring from a global perspective necessarily requires exploring how external factors compounded the legitimation crises faced by MENA regimes. The effects of globalisation at the individual level were clearly evident in the networks' youth movements built up across the world, and the manner in which they used the spread of new global communication systems. Similarly, at the state level this was evident in the spread of neoliberal economic ideas, which tied economic growth to the opening of markets and privatisation. However, there is also a broader post–Cold War security context to be considered. Since the end of the Cold War international actors have increasingly sought to erode absolute conceptions of sovereignty in favour of direct and indirect intervention. Throughout the 1990s the international community increasingly sought to construct new approaches to security, which placed humanitarianism and human security at centre stage. After the events of 11 September 2001, this discourse became a defining part of the US-led 'war on terror' to remove Saddam Hussein in 2003. Although there was no clear and imminent danger of Saddam Hussein committing genocide against his people, arguments were put forward regarding disarming Iraq of its non-existent WMD, and articulated with a humanitarian discourse based on Hussein's violent record against his people. During the Arab Spring, this humanitarian discourse was also evident in NATO's response to the Libyan crisis and the invocation of the 'responsibility to protect' (R2P). However, in the case of Libya, the Gaddafi regime was on the brink of committing genocide, allowing the international community to act swiftly to support Libyan rebels through the enforcement of a no-fly zone. Nevertheless, whilst this humanitarian discourse has been evident in the Western response to the Syrian crisis, there has been a failure to arrive at an international consensus for action; in part because the Chinese and Russian governments believe that NATO used the Libyan no-fly zone as a cover for regime change. Other international actors have come to see Western humanitarian discourses as part of a larger imperial strategy.

More indirect intervention has also resulted from Western desires to spread democracy and human rights through softer means. For example, the European Union has been a key actor in promoting its normative agenda through the European Neighbourhood Policy (ENP). This has included softer programmes of intervention in the Southern Mediterranean and the insertion of conditionality clauses in international agreements. Similarly, in the aftermath of the 11 September 2001 terrorist attacks, the United States launched its Freedom Agenda for the Middle East to promote democracy in the region. Under the George W. Bush administration, this strategy relied on promoting four pillars of political, economic, educational and women's reforms in the region. This was continued under the Obama administration, but increasingly drew on new technologies to link groups across the region and build networks of youth movements. The aim of this, through websites such as *movements.org*, was to help build 'what was already happening on a larger scale' (Hassan 2013). This strategy was not only seen as a method of combating terrorism, but also as a new method of engaging with the MENA in a period where it was clear that the status quo did not necessarily always prove accommodating to Western interests in the region. Notably, the intent behind these policies was to promote incremental reform of the region that allowed stable transitions to democracy over a period of decades, and not the fast paced collapse of traditional partners that has been characteristic of the Arab Spring. Nevertheless, this push for universal human rights and democracy is an external factor to be considered when analysing the MENA revolutions.

After the spring

Whilst the revolutions themselves brought hope of democratic transformation in the MENA, their aftermaths have left a more complicated picture. On one hand, the revolutions have undermined claims that there was a form of Arab exceptionalism that prevented the region from moving towards democracy, but on the other, concerns over the 'Islamist dilemma' have arisen along with a more complicated security environment. Whilst youth movements were behind the revolutionary impetus to remove regimes they have been far less efficient at coming together to form effective political platforms. This has allowed

Islamist parties, which were well organised before the revolutions, to reap electoral rewards, giving rise to fears that the old regimes could be replaced with 'one man, one vote, one time' systems that consolidate new Islamist authoritarian leaders. This could particularly be the case where transformations are unstable. For example in Tunisia, the revolution has led to the removal of the Ben Ali regime, the resignation of Prime Minister Ghannouchi, the dissolution of the political police, the dissolution of the ruling RDC party, and the release of political prisoners. However, the elections to the Constituent Assembly, on 23 October 2011, saw a victory for the moderate Islamist Ennahda Movement, who formed a coalition with the Congress for the Republic and Ettakatol. Since then, the assembly has been plagued by splits in coalition parties, but overall looked stable until the assassination of Chokri Belaid, the leader of the Unified Democratic Nationalist Party in February 2013. As opposition leader he had been critical of the Islamist-led government and of violent Islamists. His death has cast a shadow on the transition process, leading to violence and protests on the streets of Tunis.

In Egypt, whilst since January 2011 the Supreme Council of the Armed Forces (SCAF) has taken control of the country, they have allowed for constitutional reforms and elections. Consequently, in November 2011 to January 2012, elections for the Egyptian People's Assembly were held, which saw a sweeping victory for Islamist parties. By June 2012 these elections were deemed illegal and the Supreme Constitutional Court dissolved the parliament. In the Egyptian Presidential elections, finishing that same month, Mohammed Morsi, of the Muslim Brotherhood's Freedom and Justice Party was elected by a slim margin. A month later, President Morsi reversed the decision to dissolve parliament, and moved forward with drafting a new constitution, which was passed by referendum in December 2012. At a superficial level, this sets a positive direction for the country. However, the situation in Egypt has proved highly volatile with conflicts between security forces and ongoing protests, clashes between security services and Coptic Christians, tensions emerging between Egypt and Israel, and confusion over why two Iranian warships were permitted to sail up the Suez Canal. Moreover, the significant electoral gains made by Islamist parties in the Egyptian People's Assembly signify that the geopolitical and geostrategic orientation of Egypt is uncertain. Egypt is at the crossroads of a constitutional crisis as the SCAF and the civilian leadership seek to settle who is in charge of Egypt's destiny.

Whilst elections in Tunisia and Egypt have demonstrated the electoral appeal of Islamist groupings, Libya has bucked the trend. In July 2012, the largely secular National Forces Alliance received 48% of the vote for a new General National Congress (GNC) whilst the Islamist Justice and Construction Party received only 10%. Subsequently, the liberal independent Ali Zeidan was voted by the GNC to become Prime Minister, replacing Mustafa Abu Shagur and beating Islamist candidates. This has lessened concerns regarding the Islamist dilemma, but Libya still faces serious security challenges. Since the end of the conflict many militias have not disbanded, which has thwarted the interim government's attempts to build institutions in the country. In particular, the national priority of establishing an army has been deeply problematic, which has led to thousands of local militias becoming the dominant providers of security across the country. In turn this has allowed other security risks to emerge, which was most evident on the international stage with the murder of the US Ambassador Chris Stevens in Benghazi. Al-Qaeda in the Islamic Maghreb have used the instability in Libya as an opportunity to promote its terrorist agenda, which was witnessed across the border in Algeria when terrorists seized the Amenas Gas plant. As such, the Arab Spring, whilst confident in its move towards the democratisation of the region, has allowed new security challenges to emerge alongside non-traditional partners.

The social sciences and the way forward

The multi-causal explanation of the Arab Spring provided here demonstrates the importance of the social sciences when analysing complex issues of social transformation. To understand why these events occurred, it is necessary to look at demographic factors, along with technological innovations and the plurality of information sources to understand how a revolutionary consciousness and strategies were developed. However, analysing the level of agents alone does not provide a complete understanding, as agents work within structures. Thus, it is necessary to look at the state level, and the manner in which poor economic and security decisions contributed to alienation, poverty and a loss of dignity. Such factors helped ferment the onset of legitimation crises. Moreover, it is clear that the institutional nature of militaries played a defining role in the success or failure of the revolutions themselves, with the Egyptian and Tunisian forces stepping aside, leading to the removal of their regimes, whilst the Libyan and Syria militaries chose an alternative violent path. Yet, just as it is important to consider the internal dimensions within a state, these do not occur within a vacuum and states are no longer fully protected by appeals to sovereignty. The post–Cold War context, along with security concerns of the War on Terror, have given a greater impetus for outside actors to intervene directly and indirectly in the name of security, democracy and human rights. Moreover, as Libya demonstrated, it is clear that when the international community unites it can be highly effective.

Nevertheless, although the Arab Spring provides an opportunity, it has also delivered a range of new risks and security threats. It is unclear whether new democracies will emerge or the elections taking place in the region will be short-lived as new strongmen materialise. Similarly, it is unclear that a democratic peace will emerge as a result of the Arab Spring even if these transitions are headed in an affirmative direction. As Mansfield and Snyder's research demonstrated in 2005,

> Although democratisation in the Islamic world might contribute to peace in the very long run … simply renouncing these authoritarians [regimes] … is unlikely to lead to peaceful democratic consolidations. On the contrary, unleashing Islamic mass opinion through sudden democratisation could only raise the likelihood of war. All the risk factors are there: the media and civil society groups are inflammatory, as old elites and rising oppositions try to claim the mantle of Islamic or nationalist militancy. The rule of law is weak, and existing corrupt bureaucracies cannot serve a democratic administration properly. The boundaries of states are mismatched with those of nations, making any push for national self-determination fraught with peril … rising political participation leads to conflict and instability in states with weak political institutions. (2005: 13)

If Mansfield and Snyder are correct, then understanding the reasons for the Arab Spring is of fundamental importance for avoiding the onset of greater security threats and humanitarian disasters. By harnessing the interdisciplinarity of the social sciences to inform this debate in an enlightened and well-reasoned manner, social scientists can deploy a combination of philosophically reasoned methodologies, and state of the art methods to explore these problems and provide answers to these complex questions. By doing so, subsequent policy debates can be well informed and sound policy judgements can be constructed at the same time as human knowledge is developed in the hope of creating a more peaceful world.

Notes

1. We use the term Arab Spring here as a term to represent the revolutions that have swept the Middle East and North Africa since December 2010. However, the term itself is not without problems. Whilst others refer to it as the Arab Awakening, we note that in the region itself individual revolutions are referred to.

2. This resentment was highly evident from interviews with protesters in Tahrir Square who were keen to call for 'oil, bread and jobs', whilst also highlighting growing inequality as a reason for the revolution.

Bibliography

Feiler, B. (2011). *Generation Freedom*, New York: Harper Perennial.

Hassan, O. (2011). Interviews conducted in Tahrir Square and around Cairo.

Hassan, O. (2013). *Constructing America's Freedom Agenda for the Middle East*, Oxon: Routledge.

Khalil, A. (2011). *Liberation Square: Inside the Egyptian Revolution and the Rebirth of a Nation*, New York: St. Martin's Press.

LaGraffe, D. (2012). The youth bulge in Egypt: an intersection of demographics, security and the Arab Spring. *Journal of Strategic Security*, 5(2), 65–80.

Mansfield, E. D. and Snyder, J. L. (2005). *Electing to Fight: Why Emerging Democracies go to War*, London: MIT Press.

Noueihed, L. and Warren, A. (2012). *The Battle for the Arab Spring: Revolution, Counter-Revolution and the Making of a New Era*, New Haven: Yale University Press.

Owen, R. (2012). *The Rise and Fall of Arab Presidents for Life*, Boston: Harvard University Press.

Roudi, F. (2011). Youth population and employment in the Middle East and North Africa: opportunity or challenge? United Nations expert group meeting on adolescents, youth and development, 21–22 July. Available at http://www.un.org/esa/population/meetings/egm-adolescents/p06_roudi.pdf, accessed 23 November 2011.

US Census Bureau (2010). US Census Bureau International Data Base. Available at http://www.census.gov/population/international/data/idb/informationGateway.php, accessed 4 December 2011.

11
International Migration

Cathy McIlwaine

Introduction

A decade and a half ago, some of the most renowned names in international migration research proclaimed that, 'Like many birds, but unlike most other animals, humans are a migratory species. Indeed, migration is as old as humanity itself' (Massey et al., 1998: 1). They continued, 'Human migration is rooted in specific historical conditions that define a particular social and economic context' (1998: 3). This provides some signposts as to why the social sciences have so much to offer the study of international migration, past and present. International migration of people is fundamental not only to humanity, as they suggest, but also to the creation of nations and societies stretching back centuries. Yet its importance lies not only in its historical importance, but also in its increasing influence over the ways the global economy and societies function, as the scale of movement of people around the world continues to grow exponentially, despite more recent sluggish growth. Coupled with the increasing complexity of these movements and recognition that migration affects many more than those who actually move, research on the nature of these international flows from a wide range of disciplinary perspectives has burgeoned. The social sciences are ideally placed to understand the complexities of these movements in ways that other sets of disciplines are not.

In light of this complexity, it is essential to have a multidimensional approach in order to understand international migration. The social sciences can provide such a perspective, in part because of the combination of disciplines that engage with the study of international migration from beneath the umbrella of putative migration studies. Therefore, international migration can be viewed from a wide range of disciplinary perspectives, from what is arguably its core in geography and demography to anthropology, sociology, social psychology, political science and economics that each stress variously different scales of analysis, methodologies and theoretical standpoints. This chapter will explore a range of general issues pertinent to understanding the dynamics of international migration in spatial and temporal perspective and illustrate how a social science approach is essential in understanding one of the most important yet the most contentious phenomenon of our times. In particular, it highlights the importance of examining international migration in theoretical and empirical research terms, but also the need to use such inter-disciplinary research to feed into and challenge what are often negative public perceptions of such movement in both destination and source countries.

Delineating the temporalities and spatialities of international migration flows

Exploring the dynamics of international migration requires a relational approach which is most helpfully provided by a combination of a geographical perspective that facilitates an analysis of the global spatialities of migration and also incorporates recognition of the power relations underpinning these movements, together with a temporal approach that in part entails some form of historical analysis.

In terms of the temporalities, although international migration is at the very core of humanity, stretching back to its origins in the Rift Valley of Africa, it grew rapidly from the fifteenth century onwards as an inherent part of colonialism and industrialisation. Central to this was the movement of around 12–15 million people as slaves from West Africa to the New World, as well as the subsequent migration of indentured labourers from China, India and Japan to work on European-owned plantations (Cohen, 2008). The rise of the United States brought about the next major wave of global migration, with most between the 1860s and 1920s, with further movements within Western Europe, especially among the Irish, the Poles and the Italians. Post–Second World War economic expansion led to large-scale labour migration within Europe, together with other major movements from the UK to Australia, to Israel from across the world and within the sub-continent after the partition of India (Koser, 2009). Despite a slow-down with the oil crises in the 1970s, global migrant stock grew from 75 million in 1965 reaching between 140–155 million people by the end of the 1990s (Castles, 2000: 274–275).

This growth has continued apace in contemporary times and is manifested in multi-scalar ways. By 2010, there were 214 million international migrants in the world, up from 191 million in 2005, highlighting how such movement is on an upward trajectory despite some recent slowdown linked with the global economic recession (IOM, 2011: 49). In terms of the broad global patterns, the number of international migrants in the Global North grew by 46 million (56%) while those in the Global South grew by 13 million (18%) (UN, 2012: 3–4). This is a stark reminder of the ever-increasing pattern of uneven global development whereby the dominant global flows continue to the wealthiest and most developed countries of the world. Indeed, the US has long had the largest stock of migrants in the world, standing at around 43 million in 2010, followed by the Russian Federation and Germany, with Canada, France, the UK, Spain and the Ukraine also among the top ten in terms of actual numbers (IOM, 2010: 115). However, Saudi Arabia and India also figure among the top ten destinations, highlighting that South–South flows are diversifying. Indeed, southern countries are the origin for 70% of all international migrants, about half live in other developing nations, with South–South migration flows now equivalent to North–South (around 73 million) (UN, 2012: 1–2). Nonetheless, the growth of the international migrant population in the Global North was still mainly due to the growth in migration from the South. For example, between 1990 and 2010, international migrants in the North who had been born in the South almost doubled from 40 million to 74 million, comprising 75% of the increase in immigration in the North as a whole (UN, 2012: 4). Linked with this has been a decline in South–South migration from 39% of global migration in 1990 to 34% in 2010 (2012: 5).

While these patterns refer to migrant stocks, some interesting patterns emerge in relation to the proportion of migrants as a percentage of the total residing in individual countries. At one extreme is Qatar where migrants comprised 86.5% of the total population in 2010 (IOM, 2010: 114), while a more common pattern can be found in the UK where the share of foreign-born people in the total population stood at 12.3% in 2011 (although this also reflected a 50% increase between 1993 and 2011) (Renzo and Vargas-Silva, 2012). With cities being the primary destination for many international migrants, there are 25 cities in the world with more than 25% of their populations being foreign-born (IOM, 2010: 116). Returning to the UK case again, 42% of inner London's population in 2011 was foreign-born (Renzo and Vargas-Silva, 2012).

While this provides an overall picture of the contemporary spatialities of international migration, migration flows are also highly differentiated in terms of who moves in relation to social positioning. While the gender composition of these flows has remained broadly steady in recent years with 51% being male (IOM, 2010: 117), it is important to remember that there has been a marked 'feminisation of international migration' since the 1970s as women have increasingly moved independently rather than as so-called trailing spouses (McIlwaine, 2010). Furthermore, there have been regional and country variations in these patterns. In Europe, for example, women migrants comprise 52.3% of migrant stock with up to 57.3% in Eastern Europe and 49% in Western Europe (IOM, 2010: 184). In Latin America, there has been a feminisation of migration since 1960 when only 44.2% of migrants were women, growing to 50.1% in 2010 (2010: 154). In Asia as a whole, where women constitute 48% of international migrants, they only comprise 44.6% in South Central Asia and 55% in East Asia (2010: 165). In some individual countries, this proportion is considerably higher. For instance, in 2007, women made up 79% of migrant workers leaving Indonesia, 72% from the Philippines and 64% from Sri Lanka (UN DESA, 2009: 8). In terms of age, and while it has always been younger people who have migrated, the most notable change in recent years has been the rise in independent migration of children and young people across borders. Although most migration of people aged between thirteen and seventeen years of age is still within countries, there is some evidence that it is becoming more international in nature and that it is viewed positively by the young migrants involved, rather than assumed to be part of trafficking networks (Whitehead and Sward, 2008).

Other forms of differentiation lie in the huge diversity of different types of international migrants variously influenced by the temporalities of migration. While ostensibly, international migration refers to crossing international borders for a certain minimum period of time (usually defined as one year) (Koser, 2009), the realities of global movements is much more complex. The emergent categorisations that have attempted to simplify this complexity have often been highly contentious politically. Indeed, international migration has become increasingly politicised, linked with the fact that it tends to be viewed as a deviation from the norm, and so concomitantly is perceived as problematic because it can result in unpredictable changes in both home and host countries. Therefore, the categorisation of different types of international migrants is also linked with control by nation-states (Castles and Miller, 2003). Common types of categorisations include temporary labour migrants who migrate for short periods of time to work with often draconian restrictions on their movement; highly skilled migrants who move much more freely to work in the increasingly globalised economy; irregular migrants who are also referred to as illegal or undocumented and who enter and/or remain in a country without the legal right to do so; refugees who migrate because of fear or persecution and are protected by the 1951 United Nations Convention relating to the Status of Refugees; asylum seekers who seek protection, but may not fulfil the strict criteria laid down by the 1951 Convention; and forced migrants who may be refugees and asylum-seekers but also including those who have no choice but to move because of environmental catastrophes or ill-thought through development projects (Castles, 2000).

To add further to this complexity is a range of types determined by different motivations and temporalities. These include return migration that not only focuses on first generation migrants who have gone home on retirement or on the fulfilment of their dreams, but also the migration of what has been called 'next generations' (children and grandchildren of the first generation). Significant work has been conducted on these next generations, especially in the Caribbean context in relation to their capacity for generating socio-economic change (Conway and Potter, 2007). Various types of career migration are also important here. While this usually involves the movement of the professional classes and entailing inter-company transfer of elites, it can also include so-called shuttle migrants who make short-term border crossings to work in low-skilled jobs in construction, trading or agriculture. To this, we can also add student migration which is especially intense within Europe, but which has important international dimensions, as well as such groups as Northern Europeans who spend some time during the year in warmer climates in the south of the region.

All these categorisations and differentiations in geographical and temporal flows are underpinned by deep-seated transformation in societies in relation to globalisation and uneven development processes and the increasingly politicised nature of international migration. The need to capture such diversity and complexity has prompted the emergence of the 'new mobilities paradigm' marking a shift from geographies to mobilities. Pioneered by sociologist John Urry (2007), this approach emphasises how movements of people intersect with objects and ideas across different spaces and places ranging from the body, through transport to that of global capital and labour. Although there are obvious dangers in such a wide-ranging remit, a mobilities perspective has been especially important in understanding the fluid formation of transnational and diasporic identities, especially among feminist researchers (Cresswell, 2011) as well as the multi-scalar nature of international migration and the growing intensification of different geographies of movement. Indeed, in their attempt to move the mobilities paradigm forward, Glick Schiller and Salazar (2013) develop the notion of 'regimes of mobility' to critique the dichotomy between mobility and immobility as well as to encompass the relationships between the privileged and the exploited in their quest to move around the world.

These delineations that recognise the spatialities and temporalities of international migration in all its forms have been the result of the interchanges among social scientists in particularly fruitful ways with additional insights from the humanities contributing to the mobilities paradigm as well. Such richness in interpretation has also contributed to the theoretical frameworks that have emerged to explain international migration.

Why migrate across the globe? Theoretical perspectives on international migration

While it is not the intention here to rehearse the wide range of theoretical approaches that have been developed in order to understand these migration flows, a social science perspective has been central to understanding the forces that underpin international migration. These can usefully be identified as the structural forces in the Global South that encourage migration, the structural forces that induce migrants to move to the Global North or other regions where there are opportunities, the agency of migrants themselves, as well as the structures that arise to connect areas (Massey et al., 1998). These forces have been examined and used to explain international migration from various theoretical perspectives over time. Indeed, as migration theory has evolved, different social scientific disciplinary approaches have contributed to an ever-growing sophistication of such frameworks. While neoclassical economic theory drawing on work on rural-urban migration identifies individuals' attempts to maximise their income and wages by moving from poor to industrialised countries has been widely critiqued, economics still undergirds much theoretical work. For instance, while the New Economics of Labour Migration (NELM) viewpoint argued for a more sophisticated interpretation of migration as a risk-sharing strategy at the household level, the primary motivations of such mechanisms were thought to be income maximisation and other contextual factors such as availability of capital (Stark, 1991). However, more recent research within the social sciences has usefully refined this type of approach in terms of recognising intra-household gendered power relations in determining the nature of migration (McIlwaine, 2010). As the individual or behavioural approaches focusing on labour alone have continued to be critiqued, the examination of the role of migrant networks inherent in migration strategies has expanded considerably through structuration perspectives (Goss and Lindqvist, 1995).

As part of the recognition that the creation and maintenance of networks between places of origin and destination can produce stable migration systems, research has been further elaborated from a transnational perspective. Emerging as a critique of assimilation models of migrant adaptation (see below) rather than as a theory of the causes of migration, transnationalism has been identified by Basch et al., (1994: 7) in their seminal work as 'the processes by which immigrants forge and sustain multi-stranded social relations that link together their societies of origin and settlement'. In turn, transmigrants are those who live across borders and engage in transnational activities, contributing to the formation of transnational social fields, circuits, communities or spaces (Faist, 2000), while transnational migration has been viewed as occurring within fluid social spaces that are multi-sited, multi-local, and multi-layered (Levitt and Jaworsky, 2007). Related to this, transnational practices are viewed as fluid and diverse rather than bounded and uniform and 'transnational communities' as heterogeneous, especially on grounds of gender and ethnicity (McIlwaine, 2010, 2012). Processes of transnationalisation can be economic, social, political and cultural. For example, political transnationalisation is associated with various forms and intersections of citizenship nationally, transnationally and multiculturally while cultural transnationalisation is underpinned by processes of 'acculturation, cultural retention and strengthening of transnationally induced syncretism' (Faist, 2000: 210). Although transnationalism and its concomitant dimensions and actors have been widely contested on grounds of its reification and vagueness, it remains an enduringly diverse concept. The recognition of such complexity of forces within processes of transnationalism has arguably only been achieved because of the inter-disciplinary insights that successive and intersecting work within the social sciences have brought to bear.

A specific example of how different social scientific disciplinary insights have come together to promote understanding of international migration has been the debate on the intersections and overlaps between transnationality and diaspora. While a strict definition of diaspora refers to those who have experienced collective trauma and displacement and who live in exile, such as Jews, Africans, Palestinians, Armenians, and Kurds and who strive to return and maintain ties (Cohen, 2008), research on transnationalism has arguably led a more enriched interpretation. As Tölölian (1991: 4–5) states: 'Diasporas are the exemplary communities of the transnational moment ... the term that once described Jewish, Greek, and Armenian dispersion now shares meanings with a larger semantic domain that includes words like immigrant, expatriate, refugees, guest workers, exile community, overseas community, ethnic community.' Although, it remains acknowledged that a diaspora is not a transnational community per se, similarly it is recognised that there may be diasporic elements within the community.

Beyond tracing the dynamics and causes of international migration in its multidimensional forms, another crucial dimension of social science research has been to examine the effects of such movements on both the home and host societies at different scales and acknowledging the transnational linkages. Despite a tendency to focus on the outcomes for destination countries, increasingly the somewhat ambiguous ramifications for source nations have been articulated.

The outcomes of international migration for the Global South: the migration–development nexus

The current debates on the migration–development nexus have highlighted not only the relevance of international migration for global development policy, but also the need to further challenge the fixities of how we view the world around us. From a primarily economic perspective international migration and the remittances associated with it have been hailed as the panacea for solving development problems in the Global South. With remittances now outstripping Overseas Development Assistance in terms of amounts sent from North to South, a host of international development organisations, among them the OECD and the World Bank, have focused on the potential for remittances to reduce poverty through creating 'third way financing' of development as part of a wider neo-liberal logic (Datta, 2009: 113). When accompanied by return migration and a reinvestment of human and financial capital in home countries, it is widely thought that international migration has the potential for transforming the poor economies of the world.

However, as Alejandro Portes (2009: 7) notes: 'the change potential of migration does not always yield effects conducive or congruent with developmental goals'. This view is in line with a more circumspect interpretation that has highlighted the need to take a more socially informed view that highlights the wellbeing of migrants and the ramifications of migration on the social relations of origin and destination countries (Piper, 2009). In particular, research has shown that international migrants make considerable economic, personal and emotional sacrifices in order to send money home, although this obviously depends on the type of migrant. This can create a range of exclusions for migrants, especially financial, while the governments of home and/or host countries are absolved of responsibility for taking care of their citizens or the migrants in their midst (Datta et al., 2007). In addition, remittances are spent on a range of productive and non-productive goods with varying effects on the development of a home community, not all of which are positive and which can lead to a range of dependencies. Furthermore, the effects of 'brain drain', 'brain circulation' and 'brain gain' are also variable and can lead to depopulated towns and the emergence of various social problems linked with neglect, disillusion and fragmentation when only certain family members are left behind (Datta, 2011; Portes, 2009). In turn, although the growing role of diasporic migrant organisations and the emergence of community-based or collective remittances in the form of donations and various types of philanthropy has been acknowledged as important (McIlwaine, 2007), those that focus primarily on home areas are not always sustainable or significant in terms of actual resources invested (Portes, 2009). On a more conceptual note, some studies have shown how these debates call into question the binary conceptualisations of the Global North and South, suggesting that such a division needs to be challenged (Raghuram, 2009). Whatever the case, international migration is fundamentally re-shaping North–South relations in contradictory ways as the social and economic ties between them become ever more complex (Datta, 2009).

The outcomes of international migration for the Global North: assimilation, multiculturalism or integration?

With burgeoning flows of international migrants, their accommodation into the societies to which they move has prompted widespread intellectual, theoretical and policy responses that have been positive, negative and contradictory. Despite a long history of struggling with the governance of peoples from different cultural and linguistic backgrounds, these issues have received most attention in the second half of the twentieth century in line with the increase in international migration flows. As the classic 'Melting Pot' assimilation ideas of the US in the first half of the twentieth century were critiqued, so the importance of cultural pluralism grew since the end of the 1960s. Subsequently, notions of multiculturalism emerged focusing on the accommodation of diversity revolving around the assertion of rights among excluded minority groups and the affirmation of cultural difference, as well as economic and political participation, especially in relation to citizenship (Schuster and Solomos, 2002). Yet, notions of assimilation and multiculturalism have been highly contested with the result that different countries have interpreted assimilation and multiculturalism in varied ways both conceptually and in terms of the policies they have implemented around issues of equal rights, recognition, and citizenship with respect to migrants (Joppke, 1996). Therefore, France has emphasised assimilation of migrants, while Germany has focused on temporary guest workers who are expected to return to their home countries. The UK's position is a more fluid but potentially contradictory viewpoint that has combined reducing racial tensions and encouraging cultural diversity, while promoting the notion of British national unity.

More recently, multiculturalism has been questioned in the UK and beyond. Although some argue that multiculturalism and assimilation are not necessarily mutually exclusive, there has been a marked shift since the turn of the century towards integration. Critics have argued that multiculturalism reifies communities, ignores internal tensions and diversity within groups, and gives too much power to ethnic leaders. This has also been generated by concerns that pluralism has gone too far and that community cohesion has been threatened, as well as by fears over home-grown terrorism. Instead, migrants in countries such as the UK are now expected to integrate through a range of English language requirements, citizenship tests and ceremonies, as the migration regime makes it ever more difficult for non-EU migrants in particular to enter and settle. Although some suggest that this signals a return to assimilation approaches, it also reflects how increasing diversification of migration in countries such as the UK require integration policies to be reworked (Vertovec, 2010).

Also central to more recent understanding of how migrants settle in host societies is the relationships between transnationalism and integration. These are now recognised as mutually reinforcing processes with high levels of interaction among migrants between home and destination countries not necessarily signifying that they are less integrated, although this varies according to the nature of transnational engagement and the integration philosophies and policies in destination countries.

Irregular migration and barriers to movement

The success or otherwise of integration processes is fundamentally affected by the immigration status of migrants and especially whether they are legal or illegal, regular or irregular, documented or undocumented. This status together with the extent to which migrants integrate or not into destination societies has underpinned public debates about the importance of international migration (see below). Although highly skilled global elites can migrate with apparent ease, unskilled migrants from the poorest nations face multiple barriers to their movement except when their cheap labour is required. This has created a growing population of irregular migrants who do not have the legal right to live or work in the countries to which they have migrated. While the US has traditionally been a major destination for irregular migrants, increasingly punitive immigration legislation and the more recent securitisation of its borders following 9/11, has meant that such migration to Europe has grown and diversified. Yet similar processes of inclusion and exclusion at the EU's external borders have created so-called Fortress Europe that keeps out 'undesirable' migrants and admits the 'desirable'. The general shift towards greater immigration control has been in response to political and public opinion that overwhelmingly views irregular migration as a major threat to the societies and economies of member states. This has led to a situation whereby contemporary manifestations of international migration are inherently contradictory. The so-called liberal paradox has led to the imposition of draconian and selective immigration controls that have generated a 'global hierarchy of mobility' (Bauman, 1998).

On top of the increased bureaucratisation of international migratory movements, there has been an encroaching securitisation and militarisation of borders following 9/11, and the Madrid and London bombings, compounding the criminalisation of irregular migrants and asylum-seekers who also become potential terrorists. This is set against the fact that European economies require this labour in order to function. However, it is in the interests of capital to maintain this labour as legally vulnerable as possible, creating what has been termed a 'migrant division of labour' (Wills et al., 2010) or a 'global reserve army of labour' (De Giorgi, 2010). Indeed, the selective opening of borders has created boundaries between the 'deserving' or 'desirable' and the 'undeserving' or 'undesirable' 'others'. Borders have multiplied and become much more flexible, shifting from serving as frontiers that protect sovereignty and non-citizens from entering a nation to mechanisms that assert their power within countries and arguably maintain migrants in a continually subordinate position.

Also central to understanding the nature of migrant irregularity is the ways in which migrants create and sustain transnational identities and practices through the formation of transnational social spaces or social fields. Indeed, it has been acknowledged that transnational lives can provide sustenance and an escape from exclusion for irregular migrants (McIlwaine, 2012). This exclusion is especially marked in societies where there is open hostility of migrants and a deep misunderstanding of their position in society. Indeed, despite media portrayals of migrants making disproportionate claims on the state, research in the social sciences has highlighted how irregular migrants in particular often make significant contributions to the state's coffers through using false papers (Wills et al., 2010).

The importance of the social sciences in shaping public opinion on international migration

Leading on from this, the social sciences play an exceptionally important role in presenting robust, independent research that counters inflammatory and unfounded representations of international migrants. Indeed, it is now widely acknowledged that public opinion surveys consistently overestimate the absolute numbers of migrants in any country or region, especially for irregular migrants (IOM, 2011: xiv). A recent study of eight migrant-receiving countries (Canada, France, Germany, Italy, the Netherlands, Spain, the UK and the US) found that in all, the perceived estimate of the size of the migrant population was much higher than the actual size; for example, in Italy, this was estimated at 25% when in reality, migrants make up only 7% (2011: 8). Similar patterns have emerged in the case of the UK in particular. Another study has shown that 69% of the British public support reduced immigration and especially a reduction in irregular migration. Yet, respondents were most likely to identify asylum-seekers as immigrants (62%) when in reality they comprise only 4% of the total. In contrast, only 29% identified students whereas they make up 37% of all immigrants (Migration Observatory, 2011a: 2–3). The mismatch between perception and the available evidence highlights the populist nature of migration debates and the ease with which it is possible to blame migrants for the ills of any given society. The most common targets for blame are unemployment, security and lack of social cohesion (IOM, 2011).

This also begs the question as to how this mismatch has come about? Why and how have various sectors of the media and various anti-immigration lobbies been able to convince the general public of the dangers of immigration when there is plenty of evidence to the contrary?

There are several answers to these questions that revolve around the role of politics as well as the nature of the existing evidence base in each country. As noted earlier, immigration has become increasingly politicised in recent times and there are few political parties in sending or receiving countries that do not engage in some form with immigration in their political manoeuvrings. The politicisation of immigration debates often means that a wide range of views is not presented, which is further compounded by biased media reporting. Linked with this is the fact that the voices of key stakeholders are often ignored, such as employers whose businesses depend on migrant labour. The lack of available evidence for politicians and the media to use can also lead to distortions, not to mention uninformed policy making (IOM, 2011). In the case of the UK, the Migration Observatory (2011b) notes that the inadequacies in the evidence base means there are data gaps and limitations as well as uncertainties in the conclusions arising from academic research. This relates to problems with data sources in terms of the types of information collected or the lack of certain available data. This is especially marked in relation to irregular migration, as well as local area statistics on use of public services and housing, which are areas where perceptions are particularly divided. Importantly, they also note that the collection of more and better data will not always be possible or desirable for reasons of economic cost and/or privacy. Furthermore, better evidence will not automatically provide better migration policy solutions as the costs and benefits of migration are inherently political. In a related manner, the Migration Observatory (2011b: 15) also noted that even when the British general public identify targets for government policy, these are sometimes impossible to address. For example, low-skilled migrants are invariably identified as the most problematic yet most of this migration originates within the European Union and therefore the government has no control over these flows. Likewise, asylum-seekers are often identified yet the UK has signed a range of international conventions that binds them to provide refuge for those in need of protection.

From a different perspective, the voices of migrants are often muted in these types of debates. Not only is it rare to ask international migrants their opinions, but when they are consulted they are usually very aware of the negativity that they are subjected to in the media and by the general public at large, leading to even further marginalisation. Migrants often feel a sense of injustice at having to carry out jobs that natives are unwilling to do yet being demonised in the process. However, migrants may also be very positive about their move on economic as well as social grounds. Indeed, in terms of the growing feminisation of international migration, there is some evidence, albeit not always clear-cut, that migrant women benefit from their movement in terms of shifting gender ideologies and practices in the household and beyond that afford them greater independence (McIlwaine, 2010). Also on a more positive note, there is evidence that the more that people are exposed to migrants on a one-to-one basis, the more positive they are likely to be about them (IOM, 2011). Finally, it is also worth remembering that although international migration is generally welcomed and encouraged by poor sending countries, perceptions among the public at large in these countries are not always as positive. For example, in Honduras, 63% of the population thought it was a 'very big' problem, along with 58% of those in Argentina (IOM, 2011: 8).

Conclusions

This chapter has argued for the importance of adopting a social science perspective when examining international migration in order to capture the complexity of the phenomenon in terms of the different scales and temporalities of analysis, methodologies used to examine it and the theoretical standpoints developed. Social scientists have therefore been ideally placed to make important contributions to researching and understanding international migration. Perhaps most fundamentally of all, however, is that they have provided robust and independent research that can not only influence policy making but also challenge the insidious stereotypes and misrepresentations that can so easily dominate public opinion on this incredibly important phenomenon around the world today.

Acknowledgements

I would like to thank Michael Keith for useful comments on an earlier draft of this chapter.

Bibliography

Basch, L., Glick Schiller, N. and Szanton Blanc, C. (1994). *Nations Unbound*, Amsterdam: Gordon and Breach.

190 *Cathy McIlwaine*

Bauman, Z. (1998). *Globalization: The Human Consequences*, Cambridge: Polity.
Castles, S. (2000). International migration at the beginning of the twenty-first century: global trends and issues. *International Social Science Journal*, 52(185), 269–281.
Castles, S. and Miller, M. (2003). *The Age of Migration* (3rd edn). Basingstoke: Palgrave Macmillan.
Cohen, R. (2008). *Global Diasporas: An Introduction* (2nd edn). Oxford: Routledge.
Conway, D. and Potter, R. B. (2007). Caribbean transnational return migrants as agents of change. *Geography Compass*, 1(1), 1749–8198.
Cresswell, T (2011). *Mobilities I: Catching up, Progress in Human Geography*, 35(4), 550–558.
Datta, K. (2009). Transforming south–north relations? International migration and development. *Geography Compass*, 3(1), 108–134.
Datta, K. (2011). *Migrants and Their Money*, Cambridge: Polity.
Datta, K., McIlwaine, C., Wills, J., Evans, Y., Herbert, J. and May, J. (2007). The new development finance or exploiting migrant labour? *International Development Planning Review*, 29(1), 43–67.
De Giorgi, A. (2010). Immigration control, post–Fordism, and less eligibility. *Punishment and Society*, 12(2), 147–167.
Faist, T. (2000). Transnationalization in international migration: implications for the study of citizenship and culture. *Ethnic and Racial Studies*, 23(2), 189–222.
Glick Schiller, N. and Salazar, N. B. (2013). Regimes of mobility across the globe. *Journal of Ethnic and Migration Studies*, 39(2), 183–200.
Goss, J. and Lindquist, B. (1995). Conceptualizing international labor migration: a structuration perspective. *International Migration Review*, 29(2), 317–351.
International Organization for Migration (IOM) (2010). *International Migration Report 2010*, Geneva: IOM.
International Organization for Migration (IOM) (2011). *International Migration Report 2011*, Geneva: IOM.
Joppke, C. (1996). Multiculturalism and immigration. *Theory and Society*, 25(4), 449–500.
Koser, K. (2009). Why migration matters. *Current History*, April, 108(707), 147–153.
Levitt, P. and Jaworsky, B. N. (2007). Transnational migration studies: past developments and future trends. *Annual Review of Sociology*, 33, 129–156.
Massey, D. S., Arango, J., Hugo, G., Kouaouci, A. and Pellegrino, A. (1998). *Worlds in Motion: Understanding International Migration at the End of the Millennium*, Oxford: Oxford University Press.
McIlwaine, C. (2007). From local to global to transnational civil society. *Geography Compass*, 1(6), 1252–1281.
McIlwaine, C. (2010). Migrant machismos: exploring gender ideologies and practices among Latin American migrants in London from a multi-scalar perspective. *Gender, Place and Culture*, 17(3), 281–300.
McIlwaine, C. (2012). Constructing transnational social spaces among Latin American migrants in Europe: perspectives from the UK. *Cambridge Journal of Regions, Economy and Society*, 5(2), 271–288.
Migration Observatory (2011a). *Thinking Behind the Numbers: Understanding Public Opinion on Immigration in Britain*, Oxford: COMPAS, University of Oxford.

Migration Observatory (2011b). *Top Ten Problems in the Evidence Base for Public Debate and Policy-making on Immigration in the UK*, Oxford: COMPAS, University of Oxford.

Piper, N. (2009). Guest editorial: the complex interconnections of the migration–development nexus: a social perspective. *Population, Space and Place*, 15, 93–101.

Portes, A. (2009). Migration and development: reconciling opposite views. *Ethnic and Racial Studies*, 32(1), 5–22.

Raghuram, P. (2009). Which migration, what development? Unsettling the edifice of migration and development. *Population, Space and Place*, 15, 103–117.

Renzo, C. and Vargas-Silva, C. (2012). *Briefing. Migrants in the UK: An Overview*, Oxford: Migration Observatory, University of Oxford.

Schuster, L. and Solomos, J. (2002). Race, immigration and asylum: new labour's agenda and its consequences. *Ethnicities*, 4, 267–300.

Stark, O. (1991). *The Migration of Labor*, Cambridge, Massachusetts/Oxford: Blackwell.

Tölölian, K. (1991). The nation state and its others: in lieu of a preface. *Diaspora*, 1(1), 3–7.

UN (2012). International migration and development: report of the Secretary-General. New York: UN. Available at: http://daccess-dds-ny.un.org/doc/UNDOC/GEN/N12/452/13/PDF/N1245213.pdf?OpenElement accessed 5 April 2013.

UN DESA (2009). United Nations expert group meeting on International migration and development in Asia and the Pacific. New York: United Nations Economic and Social Commission for Asia and the Pacific Population Division. Available at: http://www.un.org/esa/population/meetings/EGM_Ittmig_Asia/BGpaper_ESCAP.pdf accessed 12 April 2013.

Urry, J. (2007). *Mobilities*, Cambridge: Polity.

Vertovec, S. (2010). Towards post-multiculturalism? Changing communities, conditions and contexts of diversity. *International Social Science Journal*, 61(199), 1468–2451.

Whitehead, A. and Sward, J. (2008). Independent child migration: introducing children's perspectives. Brighton: Migration Development Research Centre Briefing No. 11, University of Sussex, Brighton.

Wills, J., Datta, K., Evans, Y., Herbert, J., May, J. and McIlwaine, C. (2010). *Global Cities at Work: New Migrant Divisions of Labour*, London: Pluto.

Index

CPSIA information can be obtained
at www.ICGtesting.com
Printed in the USA
FFOW01n2319221015
17970FF

9 781137 269911